Materials Forming, Machining and Tribology

Series Editor

J. Paulo Davim

For further volumes:
http://www.springer.com/series/11181

J. Paulo Davim
Editor

Green Manufacturing Processes and Systems

Springer

Editor
J. Paulo Davim
Department of Mechanical Engineering,
Campus Universitário de Santiago
University of Aveiro
Aveiro
Portugal

ISSN 2195-0911 ISSN 2195-092X (electronic)
ISBN 978-3-642-33791-8 ISBN 978-3-642-33792-5 (eBook)
DOI 10.1007/978-3-642-33792-5
Springer Heidelberg New York Dordrecht London

Library of Congress Control Number: 2012949380

Printed on acid-free paper

Springer is part of Springer Science+Business Media (www.springer.com)

Preface

At this moment in time, it is difficult to obtain an exact definition of green manufacturing. It is common relate green manufacturing as "*a methodology for manufacturing that minimizes waste and pollution*". In others words, green manufacturing is developing energy-efficient manufacturing processes and systems to transform materials in products with reduced emission of greenhouse gases, reduced use of toxic materials, and reduced the generation of waste. The main objective of green manufacturing is to support future generations by attaining sustainability.

The Chap. 1 of the book provides Sustainable Manufacturing Through Environmentally-Friendly Machining. Chapter 2 is dedicated to Environmentally-Friendly Machining: Vegetable Based Cutting Fluids. Chapter 3 describes environmentally-friendly joining of tubes. Chapter 4 contains information on Concepts, Methods, and Strategies for Zero-Waste in Manufacturing. Finally, Chap. 5 is dedicated to Application of Hybrid MCDM Approach for Selecting the Best Tyre Recycling Process.

The present book can be used as a research book for final undergraduate engineering course or as a topic on sustainable manufacturing at the postgraduate level. Also, this book can serve as a useful reference for academics, manufacturing and materials researchers, manufacturing, mechanical, materials, industrial and environmental engineers, professionals in green manufacturing and related industries. The interest of scientific in this book is evident for many important centers of the research, laboratories and universities, as well as industry. Therefore, it is hoped this book will inspire and enthuse others to undertake research in this field of green manufacturing.

The Editor acknowledges Springer for this opportunity and for their enthusiastic and professional support. Finally, I would like to thank all the chapter authors for their availability for this work.

Aveiro, Portugal, August 2012 J. Paulo Davim

Contents

Contributors

Luís M. Alves IDMEC, Instituto Superior Técnico, Technical University of Lisbon, Av. Rovisco Pais, 1049-001 Lisboa, Portugal

Erhan Demirbas Department of Chemistry, Gebze Institute of Technology, 41400 Gebze-Kocaeli, Turkey

Matthew J. Franchetti Department of Mechanical, Industrial and Manufacturing Engineering, University of Toledo, Toledo, OH 43606, USA, e-mail: matthew.franchetti@utoledo.edu

Domnita Fratila Machine Building Faculty, Department of Manufacturing Engineering, Technical University of Cluj-Napoca, Bd. Muncii 103-105, 400641 Cluj-Napoca, Romania, e-mail: Domnita.Fratila@tcm.utcluj.ro

K. Jayakrishna Department of Production Engineering, National Institute of Technology, Tiruchirappalli, Tiruchirappalli, Tamil Nadu 620 015, India

Emel Kuram Department of Mechanical Engineering, Gebze Institute of Technology, 41400 Gebze-Kocaeli, Turkey

Paulo A. F. Martins IDMEC, Instituto Superior Técnico, Technical University of Lisbon, Av. Rovisco Pais, 1049-001 Lisboa, Portugal, e-mail: pmartins@ist.utl.pt

Babur Ozcelik Department of Mechanical Engineering, Gebze Institute of Technology, 41400, Gebze-Kocaeli, Turkey, e-mail: ozcelik@gyte.edu.tr

Alexander Spivak Department of Mechanical, Industrial and Manufacturing Engineering, University of Toledo, Toledo, OH 43606, USA

S. Vinodh Department of Production Engineering, National Institute of Technology, Tiruchirappalli, Tiruchirappalli, Tamil Nadu 620 015, India, e-mail: vinodh_sekar82@yahoo.com

Sustainable Manufacturing Through Environmentally-Friendly Machining

Domnita Fratila

Abstract The interest in pollution prevention is continuously growing. This determines several industries, including manufacturing, to develop and implement various environmentally-friendly strategies. Product design, selection of raw materials, manufacturing process, product delivery and reuse or recycling options for products' end of life have influences for the of environmental degradation level. There is an ongoing search for new and innovative ways by which industry can lessen its impact on the environment. Efforts are currently focused to: efficient consumption of resources and conserve energy, minimize the environmental effects of energy production, improve waste management system. This chapter presents several aspects regarding the environmental impact of manufacturing process and the necessity of changed process for increasing their sustainability and thus, preventing polluting generation. It is mainly focused on investigating various aspects of machining process from an environmental perspective.

1 Introduction

The society has generally two kinds of interactions the environment: as a source for natural resources, and as a landfill for solid, gaseous and liquid wastes. The damages act as depletion and the reduced quantity and quality of resources and as unbalancing the conditions of previously natural processes. The change in balance takes usually years to detect and can be influenced by a variety of factors. This issue makes identification and isolation of the problems difficult and sometimes controversial.

D. Fratila (✉)
Department of Manufacturing Engineering, Machine Building Faculty, Technical University of Cluj-Napoca, Bd. Muncii 103-105, 400641 Cluj-Napoca, Romania
e-mail: Domnita.Fratila@tcm.utcluj.ro

J. P. Davim (ed.), *Green Manufacturing Processes and Systems*, Materials Forming, Machining and Tribology, DOI: 10.1007/978-3-642-33792-5_1,

The studies done in this direction leads to identification of several aspects concerning the environment depreciation: ozone depletion, global warming, acidification, and eutrophication. Corrective actions involve changes in the types and ways it uses materials and energy for the production, use, and disposal of products.

A life cycle assessment scheme can properly link each product or process with its environmental load. Although this task is conceptually simple, it is quite complex in fact. The major complexities originate in difficulties to: establish the system boundaries, obtain accurate data, represent the data with concise descriptors that appropriately assign responsibility, and to evaluate properly the results.

The idea of a product life cycle is generally regarded as a materials flow process that starts with extraction of raw materials and ends with the disposal of the waste products. The general stages of this linear cycle are: material extraction, primary processing and refining (premanufacture), product manufacture, product distribution, use, and final disposition. This sequence follows the principal product material flow, but of course there are multiple cross flows (secondary flows) as well as backflows (part remanufacturing, product reuse, or material recycling). For each of these stages the environmental stressors are: materials choice, energy use, solid waste, liquid residues, and gaseous residues [1].

The manufacturing processes seem to be quite benign compared to materials extraction and primary processing, but manufacturing processes set many of the requirements for primary processing outputs. Normally, the processes with higher scrap rates require more energy in primary processing, while processes which use large quantities of recycled materials will have reduced primary energy needs.

Concluding the manufacturing uses materials and energy (not directly incorporated into the product) and then eliminates them as wastes or emissions to the environment. In addition to work pieces, tools and energy, a second environmentally important category of auxiliary materials used in manufacturing processes is metalworking fluids, cleaning fluids and coatings. Lubricants and solvents are of particular concern, being used to remove the coolant or lubricants from the surface of the parts [1].

2 Sustainable Manufacturing Technologies

Sustainable manufacturing is a relatively less-known and significantly element of sustainable development, including three functional elements: sustainable products, processes, and systems. It is important to develop quantitative predictive models for sustainable product design and manufacture in order to understand the integral role of these elements of sustainability in product manufacture.

One of major challenges to the industry is to design and manufacture sustainable and environment-friendly products. Such process involves complex, interdisciplinary approaches and solutions. According to definition of sustainability, the sustainable products are fully compatible with environment throughout their life cycle. The manufacturing processes must exceed beyond their traditional

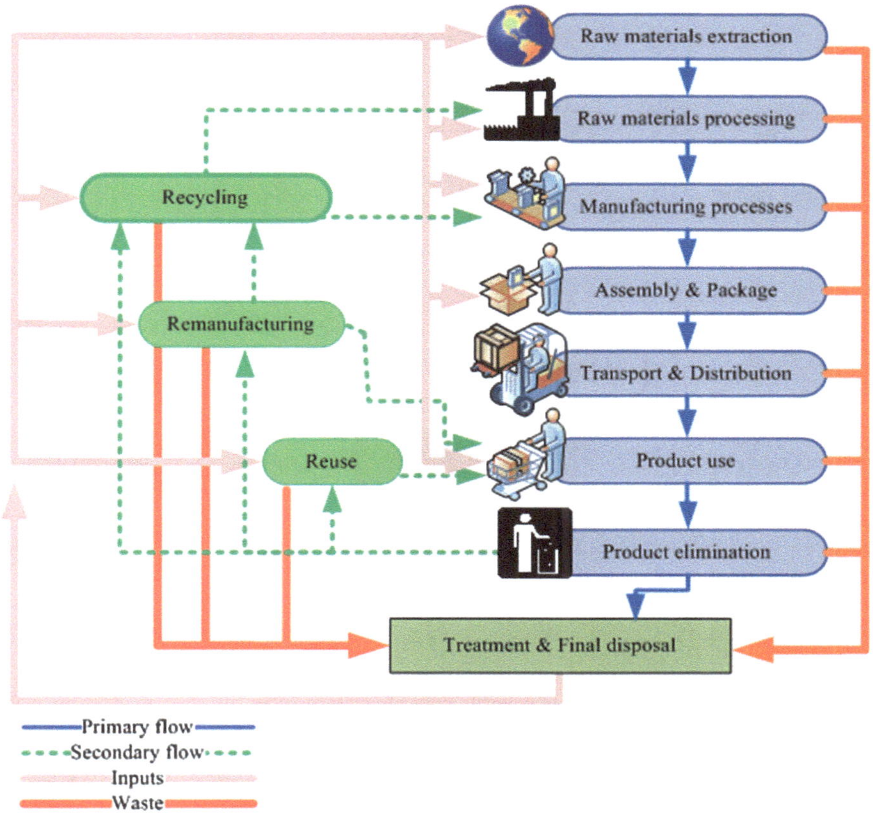

Fig. 1 Typical life cycle stages

requirements of functionality, cost, performance, and time-to-market, by considering also sustainability. This would be possible by thinking in terms of reduced material utilization, minimizing energy consumption, waste-free manufacturing processes, and resource recovery following the appropriate end of product scenario [2].

In recent times the goal of reuse, recycling, and remanufacturing emerged with innovative engineered materials, manufacturing processes, and systems aiming to provide multiple life-cycle products. The need for ecological efficiency and the environmental concerns are often associated with production of minimum waste amounts, minimum toxic emissions into the air, soil, and water, and minimum energy consumption at all life cycle stages (Fig. 1).

The companies have the potential to save costs and to improve their environmental performance even the production stays on the same size or it is decreased. This is possible with the implementation of the sustainability principles in the manufacturing processes [3– 6].

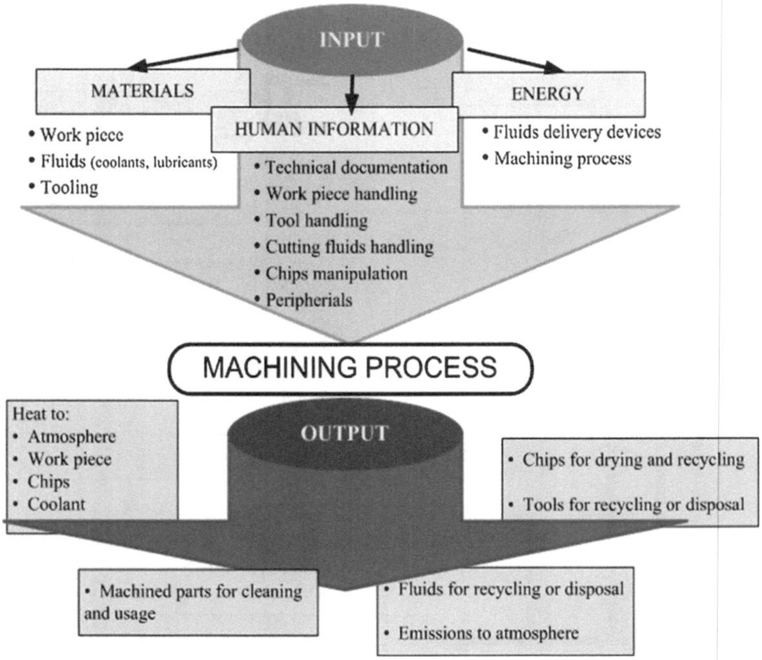

Fig. 2 Overall machining process of a part

As already known, machining is a material removal process that typically involves the cutting of metals using various cutting tools. A process that removes material, machining can be inherently wasteful due its use of raw materials (workpieces) and energy. Machining processes are particularly useful due to their high dimensional accuracy, process flexibility, and cost-effectiveness in parts' producing. Machining is unique, among manufacturing processes, in that it can be used both to fabricate products and to finish products [7].

Overall production process of a component consists of several elements and steps: process inputs (material, energy, data), machining, cooling/lubrication, part cleaning, preparation of chips to be disposed and their interactions [8–10] as shown in the Fig. 2.

Major environmental hazards in machining operations are due to the use of cutting fluids [11]. Direct exposure of the production worker to these fluids can lead to skin diseases and respiratory disorders and there is also an increased health risks [12]. The cause is attributable to both the original constituents in the fluid and impurities which are introduced or generated during operation. Losses of cutting fluid from the manufacturing process occur through vaporization, loss with chips and work pieces as they leave the machine, loss with machine components such as manipulation and transport devices, as well as losses through vacuum and air pressure systems and through droplet formation and ensuing leakage [13].

Up to 30 % of the annual cutting fluid consumption is lost through removal from the system by the above means. In addition, contamination of chips generated makes them difficult to recycle and work pieces must frequently be cleaned before proceeding to the next process step. Such cleaning operations are non-value adding and can also add to the overall environmental issues of the manufacturing system. In addition to environmental impact, it has been estimated that the use of cutting fluid can account for over 15 % of machining costs [14].

There are several ways to reduce or avoid the usage of cooling lubricating fluids (CLF). The question that occurs is what kind of CLFs and how much of each have to be used to the manufacturing processes and their costs. In the case of conventional CLFs, their purchase costs have to be increased tacking into account CLFs disposal costs, parts cleaning and drying, depreciation costs, maintenance costs and the costs connected with personnel and health issues. Dry cutting therefore results in both environmental and economic benefits.

2.1 Dry Machining and Near-Dry Machining

In normal machining operations, CLFs are used to flood the area of contact between the tool and the work piece. The most radical approach is to convert conventional flood cooling to dry cutting which eliminates all problems associated with cutting fluids. The process carried out in the absence of CLFs is called dry machining (DM), or near-dry machining (NDM) when the process runs in the presence of a very small quantity of lubricants, in the range of 10–150 ml/h, delivered in the cutting zone in a mist way. Dry cutting is made possible by recent developments in cutting tool materials, but presents new challenges in the areas of chip handling and machine tool design. Thus, the use of NDM [called also Minimal Quantity Lubrication (MQL)] is the way that can lead to a significant cost reduction of the process [15].

By simply avoiding CLF usage and applying DM alternative (using additionally new high performance coated cutting tools), there would be a huge progress for sustainable technologies [16, 17, 18, 19]. But, there are new work piece materials used especially in aerospace industry (such as nickel alloys, titanium alloys, Co-Cr alloys) which are extremely difficult to machine. In combination with high cutting speed, serious difficulties will be encountered, disabling the use of DM.

Tasks of coolants/lubricants, in the case of their absence have to be taken by other machining process components. In this respect, the latest technique concerns in demonstrating the application of modified tool for DM use [20]. A solid lubricant (molybdenum disulfide) was filled into the micro-holes on the rake or rake and flank face of the cemented carbide (WC/Co) tools in order to create self-lubricated tools. During dry-cutting of hardened steel, the cutting forces, the tool wear and the friction coefficient at the tool-chip interface using a such tool are significantly lower compared with the use of a conventional WC/Co tools. This effect is caused by the self-lubrication action of the modified tool.

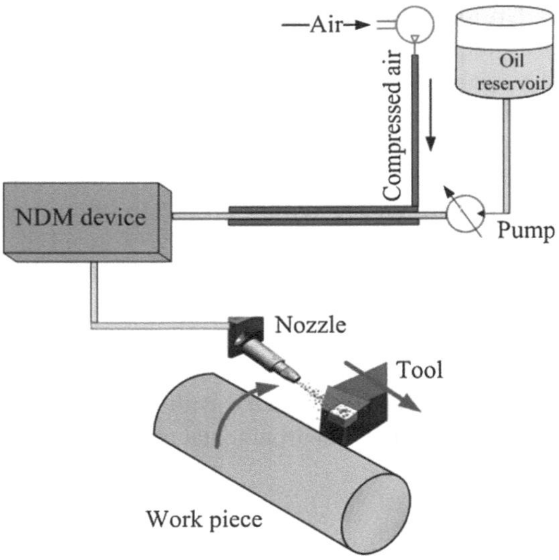

Fig. 3 Near-dry machining principle

For both technologically and economically optimal results, the NDM must be considered as a system having the following components: CLF feed technology, NDM media, parameter settings, tools and machine tools. The principle of this technique is schematically presented in Fig. 3.

In NDM the most commonly used media are synthetic esters and fatty alcohols, but some applications are still using emulsions or water (Table 1). Such high-performance oils have excellent lubricity and biodegradability properties and they are environmentally friendly [21, 22].

These fluid media are fed in very small quantity to machining area with or without the assistance of a transport medium. In the case of air-less systems a pump delivers CLF in the form of a rapid succession of precision-metered droplets. In the case of high pressure systems, the medium is atomized to form extremely fine droplets delivered to machining point in form of aerosol spray using compressed air at 4–6 bar pressure.

CLF delivery system can be an external supply with one or more nozzles fitted separately in the machine area or an internal supply of the media using the channels built into the tool body. Important roles in the quality of process results play the following elements: number, direction and emplacement of nozzles.

A distinction should be done between MQL and minimal quantity cooling (MQC) depending on the type and main functions of CLF. When good lubrication properties are needed than oils are used as cutting fluids, their function being to reduce friction and adhesion between work piece, chips and tool and thus the amount of friction heat generated. Much less frequently than oils, are used emulsions and water, when it is essential to cool more efficiently the tool or the part. These operations are regarded as MQC [22].

Table 1 General MQL fluids characteristics

Synthetic esters (chemically modified vegetable oils)	Fatty alcohols (alcohols made from natural raw materials or from mineral oils)
Toxicologically harmless low level of hazard good biodegradability	
- Very good lubrication property - Good corrosion resistance - Inferior cooling properties - Vaporizes with residuals - High flash and boiling point with low viscosity	- Poor lubrication properties - Better heat removal due to evaporation heat - Little residuals - Low flash and boiling point, comparatively high viscosity

There are several application areas for DM and MQL. It involves a wide range of material—process combinations such as drilling reaming, taping, milling turning, gear milling, sawing, broaching, grinding, respectively aluminum, aluminum alloys, high alloyed bearing steel, tempered steel, cast irons.

CLFs avoidance requires alternative solutions that can be achieved through appropriate design of machine tools. The primary functions of the cooling lubricants from flood cooling process should be overtaken from other system components or solved by other technical solutions. In order to be suitable for DM, the design changes of existing equipment could require a lot of effort and high costs. Thus, if the machine tools rebuilding is not economically feasible, the application of MQL supply system and the housing of working area are usually a reasonable and necessary changes [22, 23].

For the new developed machine tools, the possibility to make substantial design modifications is easier, since requirements are known from the beginning. There is not only a single practical solution concerning the boundary conditions, since the manufacturing equipments are so different (individual machine tools, transfer lines, machining centers etc). In mass production and flexible manufacturing systems, the demands concerning the manufacturing structures changes can be divided in several classes: chip removal from the working area, temperature compensation, MQL system integration, and safety measures [22, 24].

With the increasing trends in the achieving sustainable machining, DM and NDM are emerging as viable and sustainable alternatives to the flood cooling (FC) in the machining processes. Even NDM is slowly being accepted as an alternative to FC and provides up to 80 % longer tool life, further studies are needed to find the best design of external nozzles and getting the right amount of fluid to the cutting point.

Use of these two machining methods could offer the possibility for efficiency increasing, due to the lubricants elimination in the form of conditioned waste, the process chain optimization, and the improvement of the work environment as well. The procedure is strongly dependent on the concrete process characteristics, the material properties of work piece and cutting tool.

2.2 Cryogenic Machining

Another innovative method of cooling the cutting area (cutting tool and/or work piece) during the cutting process is the cryogenic machining (CM). This is a very effective machining technique that utilizes a cryogenic CLF as coolant to bring down the temperature at the chip-tool interface, reducing the tool wear or to change the characteristics of material and, thus, improve machining performance and product quality [20, 25]. By researches four cryogenic approaches have been attempted: work-piece pre cooling, indirect cryogenic cooling, and cryogenic jet spraying and direct cryogenic treatment of cutting tools.

Usually the coolant used is a safe, noncombustible and non-corrosive gas (nitrogen), liquefied by cooling to −196 °C. It is not usual to employ the air as coolant since all gases have relatively poor cooling properties compared with liquids. Better cooling performances have been achieved by refrigerating the gases, high pressure of gases being also considered an additional help.

Basic principle of this process and the liquid nitrogen (LN) production process are schematically presented in Fig. 4. LN returns to the atmosphere by evaporating quickly in the cryogenic machining system, it leaves no residue to contaminate the parts, chips, machine tool and operator and the coolant disposal costs are eliminated.

CM can be successfully used in machining of difficult-to-machine materials, hard materials, high abrasive materials, or super alloys [26]. The method, introduced first in industry in 2003, was specially developed for turning and some

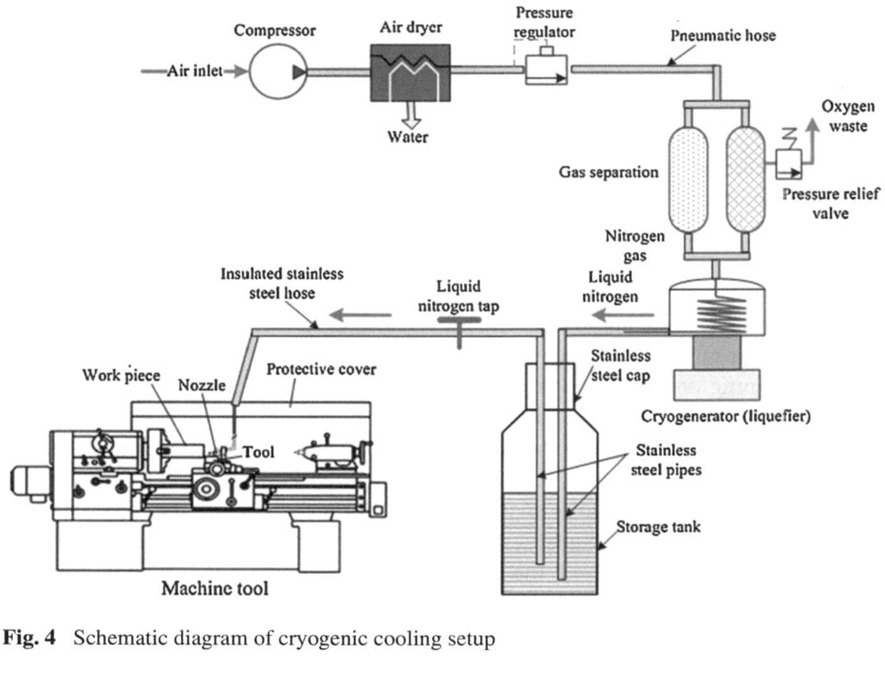

Fig. 4 Schematic diagram of cryogenic cooling setup

milling applications. Nowadays it is implemented for high speed machining for hard and super alloys in aerospace and automobile industries. The tool manufacturers produce new machine tools that utilize cryogenic machining technology [27].

Other improvements consist of a faster machining process, a higher parts' quality, a better machining performance, and an overall cost reduction. Summarizing, some of advantages of CM are: reduced chip-tool interface temperature, reduced tool wear, higher production rates, greater product quality, power saving, no mist collection, no filtration, no wet chips, no contaminated work piece or disposal costs. The main disadvantages of this technique are the additional equipment costs and the high price of LN, which is not reusable.

This technique can be considered a sustainable machining being a clean, safe and environmental-friendly method. It allows increasing the process productivity by a higher material removal rate without an increase of tool wear and with a reduced cutting tool changeover costs.

2.3 High Pressure Jet Assisted Machining

Even the trend in machining is the implementation on the large scale of DM and NDM (combined with the most advanced tool materials and coatings) in order to reduce the consumption of cutting fluids [22, 24], there are cases when the lubricants can not be completely avoided. One of the techniques developed to replace the conventional process, keeping or even increasing the machining performance, is high pressure jet-assisted machining (HPJAM).

HPJAM is another innovative method of lubricating/cooling the tool-chip interface during the cutting process, using the thermal and mechanical properties of a high-pressure jet of water or emulsion directed into the cutting zone. This technique is related to delivering the oil-based or water-based CLF under extremely high pressure in the range 80–360 MPa to the cutting tool tip through small diameter nozzles (0.15–0.25 mm) [3, 4, 24, 28]. The basic principle of HPJAM is shown in Fig. 5.

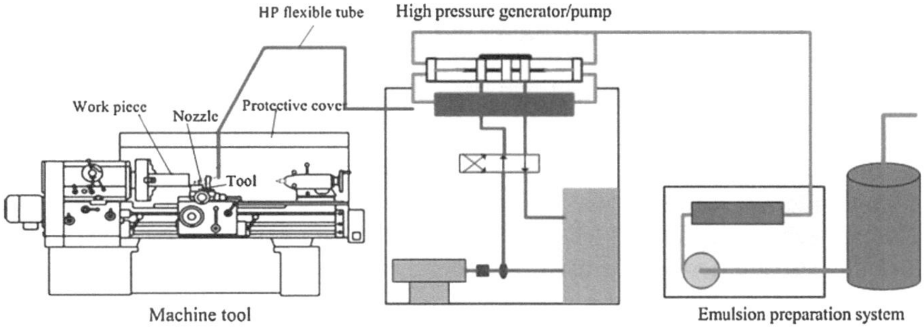

Fig. 5 Schematic diagram of high pressure jet assisted machining

Compared to conventional flood, the CLF flow rate is in this case relatively small. Due to their higher flexibility, external nozzles are widely utilized compared to internal channels. Having a high pressure, the CLF jet can penetrate closer to the share zone and cools it. The fluid jet is directed to the cutting edge at a low angle directly between the rake face and the chip.

Besides the cooling effect this method allows the control of the friction conditions between the tool rake face and the chip back side, leading to a machining performance improvement. The initial goal of HPJAM was a important increase of material removal rate and process productivity of advanced materials such as hardened steels used for moulds, Ni-based and Ti-based alloys used in gas turbines, Cr-Co alloys used for prosthesis, and in the aerospace industry [3].

Some of its disadvantages consist of higher initial capital investment for equipment and the fact that the oil-based CLFs are still used. The machine tool should be fitted with high pressure equipment. The system components involve: a high pressure pump supplied with filtered water or emulsion, high pressure pipes or tubes, a nozzle fixed beside a tool holder clamped with adjustable system, and a filtering system. By implementing of this method in the practice, the relatively larger energy consumption (which is compensated by higher removal rates and longer tool life), the noise and emulsion aerosols have to be considered.

Concerning the potential benefits of HPJAM, they refer to the sustainability through lower flow rate of CLF in comparison to conventional machining and the lower cutting forces and longer tool-life. This technique allows an extension of the operational area up to 35 % increase in both the maximum achievable feed rate and cutting speed [3, 28]. Other advantages are the significant improvement of chips' breakability and the increased process productivity by the extension of the machining parameters operational ranges. Unfortunately there is still technological gap concerning the poor investigations of the relationships between the HPJAM process parameters (that act and interact in a complex manner) and the machining performance responses.

3 Evaluation of Technologies' Sustainability

3.1 Assessment of Machining Process Sustainability

The manufacturing process is isolate from the global sustainability concept. Sustainability studies related to manufacturing processes are based on development of a practically implementable tool as manufacturing process sustainability index through literature survey and experimental work. This can be achieved in different stages starting with characterization. The data collected and the existing modeling capabilities will be used to model the impact of the manufacturing process on the main contributing sustainability parameters.

A complex modeling technique, involving analytical and numerical methods, coupled with empirical data and artificial intelligence techniques, must be developed in order to quantify scientifically the influence of each parameter. Then, the modeled production process can be optimized to achieve desired level of sustainability with respect to constraints imposed by all involved variables. These optimized results can be used to modify the existing processes and enhance the manufacturing performance with respect to the main factors considered [2].

The optimized results can be finally used in defining the sustainability rating for the specific manufacturing process. For the selected process, the weighing factors can also be used to evaluate and to serve the customized application in establishing the final sustainability rating. Two of the most-needed features of the proposed sustainability assessment systems for machining processes are the user friendliness and communication efficiency.

The main goal in identifying and defining the various elements that contribute to manufacturing process sustainability is to establish a methodology to evaluate the sustainability level of a manufacturing process. Such evaluation can be performed independently of the product life-cycle, recycling, or remanufacturability of the product that is manufactured. Requirements of sustainable manufacturing covering decision-making aspects and recycling are life-cycle assessment, environmental costs, and supply chains [2].

Manufacturing processes are numerous and differ widely, depending on the manufactured product, the fabrication method, and process parameters. Due these considerations, the identification of the factors involved in process sustainability and the demarcation of their boundaries becomes a complex issue.

Processing cost depends also on the method used to produce the part/component and the work piece material selected. In the effort to minimize the manufacturing costs, the industrial organizations endeavor to maintain the product quality, power consumption, and operator and machine safety. In the cases when the processing includes the use of coolants or lubricants and the emission of toxic and harmful materials, this poses environmental, safety, and personnel health problems. Among the various influencing factors, the following interacting factors are relevant to make a manufacturing process sustainable: (1) Manufacturing costs; (2) Environmental impact; (3) Energy consumption; (4) Operational safety and personnel health; (5) Waste management [2].

Within the manufacturing process sustainability assessment only the manufacturing costs involved during the manufacturing operation times, including the tooling costs, are considered. In addition, there are also other direct and indirect costs related to environmental consequences safety aspects and operator's health. It is needed to count the costs for recycling and reusing of consumables, chips and coolants or lubricants.

Environmental impact contributes to pollution by several basic factors such as resources' depreciation, chips, metallic dust, use of toxic, combustible and explosive materials, waste of coolants and lubricants and emissions from metal working fluids use.

In manufacturing processes the energy saving fort the entire operational period of the machine is one of most needed sustainability factors. The issue consists of monitoring the power consumption rate and evaluating energy efficiency.

Fig. 6 Factors affecting the machining process sustainability and their desired levels

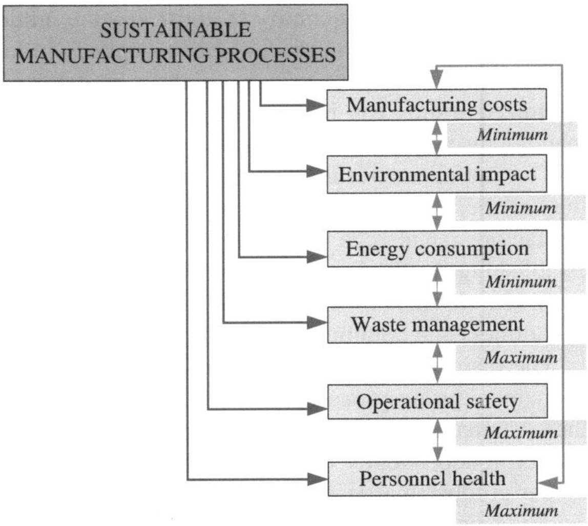

Particularly, in the machining processes the power consumption can be decreased by setting optimal cutting conditions, facilitating better tribological conditions, the use of proper lubricants and coolants, the selection of cutting tool inserts and the cutting tool-work piece material combinations. It is clearly that for sustainability assessment of energy/power consumption are preferred the environmentally friendly energy sources. The factor concerning the source of energy can be added in the rating system of process sustainability if the renewable sources are available and widely used in the industry.

Waste management category accounts the recycling and disposal of all wastes resulted during and after the manufacturing process is completed. Even the zero waste generation with no emissions in environment is the ideal process from ecological point of view it is technologically not feasible yet in the practice. Thus the efforts are focused on finding solutions in order to reduce or eliminate wastes.

The operational safety is focused on the ergonomic design of human interface and the possible unsafe interaction between operator and machine during the manufacturing operations. In relation to the manufacturing processes, the safety aspects are generally divided in two categories: personnel safety and work safety.

The last element contributing to the machining process sustainability is the personnel health. Its assessment is based on the compliance with the national and international regulatory requirements imposed to industry, and it concerns the admitted level of emissions and waste from machining operations and their impact on the exposed areas and operators. The personnel health and working area is commonly affected by exposure to the mist and vapors from metal working fluids used as coolants or lubricants during manufacturing processes. These contain usually a large amount of additives in order to enhance the process performance.

At preliminary stage of sustainability evaluation, it is not excluded the consideration of other secondary parameters such as the product's functionality requirements. They could influence the decision making process being related to energy consumption and machining costs, the marketing strategies and the initial equipment investment. There is a strong interaction between these six main factors even they have different expectation levels [29], as shown in Fig. 6.

Obviously, they can not achieve their best level due to technological reasons and cost implications. A combination of minimum and maximum of factors' levels will be involved in an optimized practical solution within the constraints imposed.

The analysis of above mentioned factors shows how the sustainability measures can be selected along an appropriate method in order to optimize the machining parameters for a high sustainability.

3.2 Assessment Methods

Before starting to use any assessment method, it is important to identify first the boundaries of the system to be examined. Particularly, in the case of machining, the overall system includes activities such as material production, tool preparation, material removal, and cleaning. Figure 7 shows a general machining scenario with the important process stages. The processes included in the diagram will be briefly examined in order to provide a rough estimation of environmental impact. The items marked by grey color have been omitted in the macro-level analysis [3, 4].

The consumption of cutting fluids is one of the critical issues related to environmental impacts of the machining process. It is already recognized that the

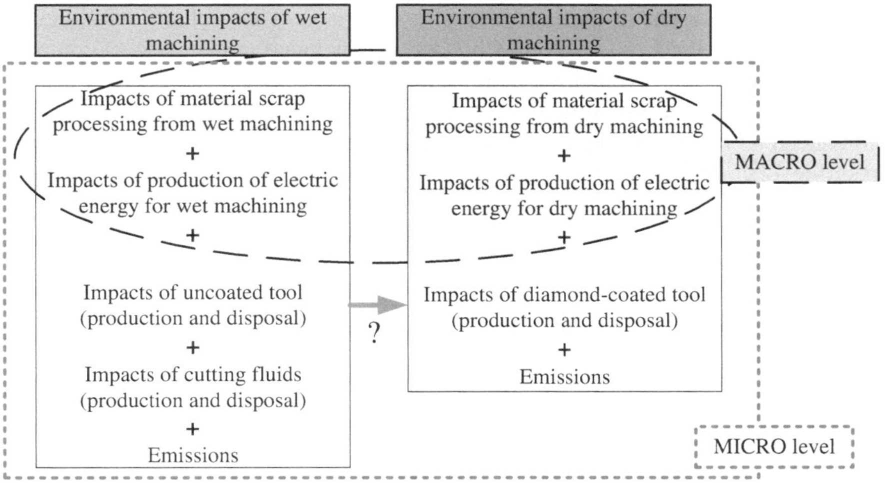

Fig. 7 General machining scenario

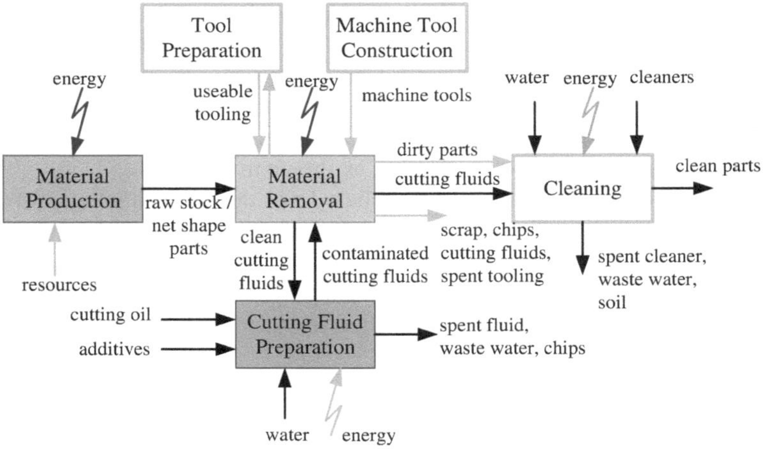

Fig. 8 Micro and Macro level comparison [4]

effect of cutting fluids on the environment, with respect to their degradation and their ultimate disposal, is a major problem. The disposal of cutting fluids at the end of life results into soil and water contamination [3, 4, 30, 31]. Also, the cutting fluid that adheres to the metal chips creates a problem for the metal recycling. The metal chips have to go through the waste processing which may include the cleaning, the separation, etc.

The list of the components required for the micro-level comparison between NDM and FM is illustrated in Fig. 8.

This also includes the components (i.e., impact of cutting tools, cutting fluids) which are not present in the macro-level comparison. The higher environmental performance of NDM process to FM process is concluded in macro-level comparison. In order to controvert this relationship in comparison at the micro-level, the environmental impacts of production, the disposal of TiN-coated tools and emissions need to be greater than the summation of environmental impacts shown in macro-level comparison.

3.2.1 Material Production

The material production is an important factor to be considered due to its environmental implications. The production of materials is energy- and resource-intensive processes. While material production seems to be outside the system boundaries of machining, the machining can be viewed as a process that pulls in the raw materials, altering them in the course of producing products.

In creating products, the machining process often uses large amounts of material. In many cases, only a fraction of the total material entering into the manufacturing plant leaves it in the form of a product. The estimations of scrap production in the machining range between 10 and 60 % [32].

While these chips and scraps can be recycled, the machining process itself requires the inflow of a large amount of pure material, raw material coming from virgin sources. The consumption of those materials from the virgin sources requires more energy than the use of materials from recycled sources requires only. This is an important process requirement that must be considered when evaluating machining [33, 34]. Thus, the importance of tracing back material flows to material production is obvious.

3.2.2 Cutting Fluid Preparation

The cutting fluids are another important part of machining, both in terms of operation and in terms of environmental impact. The most popular type of cutting fluid is soluble oil. In use, soluble oils are typically diluted with water, such that around 95 % of the cutting fluid, by volume, is water [4]. The other 5 % is a combination of oil, emulsifiers, and additives [35–38]. The additives are used: to limit the corrosion, to control the acidity, to control the microbial growth, to improve the lubricity, and to prevent the foaming.

Given the estimations of the metalworking fluids use along with a work scenario (i.e. 52 work weeks per year, with five work days per week), the values for the amount of concentrated metalworking fluid and water used per machine per day can be obtained. Once formulated, cutting fluids can be circulated through a system numerous times. However, losses frequently occur, often through vaporization or through chips, scrap, and work pieces leaving the material removal process [39]. Some estimations show about 10–30 % of the annual total cutting fluid consumption may be lost through these mechanisms [30, 32].

The cutting fluid will pick up contaminants such as metal chips, fines, and tramp oil. Such contaminants can be removed using a separation or filtration process, or the cutting fluid can be disposed of and replaced with fresh fluid. The disposal costs of spent metalworking fluid are approximately equal with the cost of the replacement fluid [40]. In the case of NDM the cutting fluids are completely consumed and, thus all these costs are avoided [41].

3.2.3 Tool Preparation

While tooling plays a major role in the machining process, the direct environmental impact of tooling is limited. Due to their relatively long life, the environmental cost of tools and tool maintenance is often amortized over numerous products. This makes their contribution to the environmental impact relatively insignificant in a per machined part analysis. The selection of appropriate tools can allow increasing the material removal rates, thereby reducing the total machining energy required [4].

Currently the most machining processes are done using carbide tools. A large proportion of these carbide tools are sold as indexing inserts, cutting inserts that attach to specially designed tool holders [42].

The carbide tools production does require some energy intensive materials and processes. Some of the manufacturing steps, including sintering (used to form the carbide tool) and physical vapor deposition (PVD) or chemical vapor deposition (CVD) (used to coat the carbide), are also quite energy intensive [43]. The fact that carbide cutting tools can be used numerous times on multiple surfaces means that this energy investment is distributed over numerous parts. Thus, the per part energy contribution from tool production can be more or less ignored, particularly in light of the material removal and material production analyses.

The alternatives to the carbide tools do exist, the most popular being high-speed steel (HSS). The HSSs are still used in the majority of the drilling applications, as well as in many milling applications [44]. Like the carbide tools, the HSS tools can also be coated through PVD or CVD processes. As mentioned earlier, perhaps the biggest difference between high-speed steel tools and carbide tools lies in the machining time [42].

3.2.4 Machine Tool Construction

The machine tools clearly play a major role in the machining process, but their direct environmental impact is limited. The most of the machine tools are in use for many years. These long lifetimes mean that the environmental impact of machine tool construction is amortized over numerous products over many years. Thus, the environmental impact per machined part is relatively small. The big effect of machine tools on machining has to do with energy efficiency. The newer machine tools can be significantly more energy-efficient than the older machine tools, resulting in the energy savings during material removal. The efficiency improvements could reduce the energy requirements per unit of material volume removed by approximately 50 % [3, 4].

3.2.5 Material Removal

Most of the environmental impact from the material removal process stems from the energy use. In estimating the energy requirements for material removal, the specific cutting energies are often used [4]. While the cutting energies for machining can depend on many factors, including the work piece properties, the presence of cutting fluids, the sharpness of cutting tools, and the process variables, the ranges of approximate cutting energies in machining are available [45]. Determination of specific cutting energies can help to calculate the minimum amount of energy required to machine a certain volume of material.

In the machining processes, in addition to providing energy to the tool tip, additional energy must be provided to auxiliary equipment such as the cutting fluid feeding equipment, the work piece handling equipment, the chips handling equipment, the tool changers, computers, and machine lubrication systems. When the auxiliary equipment is present, the energy needed of the auxiliary equipment can far exceed the actual cutting energy requirements. The energy use per amount of material removed can be estimated by following data [3, 4, 46]:

- *Energy Breakdown* shows how total energy use is distributed among various activities;
- *Constant start-up operations* refer to start-up energy use, such as for computers and unloaded motors;
- *Run-time operations* include energy used to position materials and load tools.
- *Material removal operations* refer to the actual energy involved in the machining.

Energy Use per 1,000 *work hours* can be calculated using: the number of hours spent powered up but idle, the number of hours spent positioning and loading, the number of hours spent actually removing material, the energy required to run the machine while idle, the energy required to run the machine while positioning and loading, and the energy required to run the machine while removing material.

Material removed per 1,000 *work hours* can be obtained by estimating a material removal rate. This estimation is difficult, as material removal rates depend on numerous parameters, including tool material (HSS versus carbide), part material (aluminum versus steel), part design (fine versus rough geometry), and processing parameters (flood versus dry machining).

With energy and material removal data for each machine, the amount of energy required per amount of material removed can be calculated. These values provide a general estimate of the energy requirements for material removal operations in machining. The values should provide a good order-of-magnitude estimate of the energy requirements for the material removal process.

3.2.6 Cleaning Process

The cleaning plays also a role in the machining, being one of the most often mentioned when discussing environmental impact. The importance of cleaning, and the environmental impact of cleaning, is highly dependent on the product. By using NDM techniques the cleaning of the products are not necessary anymore. This highly diversified cleaning landscape, both in terms of amount of cleaning and type of cleaning, make general qualitative analysis of this process difficult.

Metal cleaning was dominated by several large-use chemicals that could be used in a wide array of different situations. Currently the numerous different cleaning solutions have been implemented. Many of the new cleaning processes rely on aqueous cleaners instead of solvent cleaners [47, 48].

3.2.7 Environmental Concerns

The analysis of machining presented above, and particularly the analyses of the material removal and material preparation processes, focus heavily on energy use. Energy use and energy sources are important to examine when investigating environmental impacts [4].

While the environmental concerns associated with material removal and material production are focused on energy use, the environmental aspects associated with the cutting fluid preparation and cleaning are more closely to liquid and hazardous waste. These pollutants raise issues at both local and global levels. While some of the chemicals used in these processes can be harmful to workers, such as some additives to cutting fluids, other chemicals are associated with the high-level ozone depletion [4].

4 Conclusion

Sustainable production became a global concept, contenting important elements on all the fields as well as machining processes. Due to the waste and related emissions from the primary processes, metals processing is considered an important source of environmental damages. The way to sustainable manufacturing through environmentally-friendly machining begins from the steps that must be taken to implement ecological machining methods in order to make these technologies reliable, environmentally friendly and cost efficient.

Many elements and aspects of manufacturing processes have important implications for the environment state, but many products can be manufactured by more alternative processes. Often, one of them involves the use of less damaging substances of than the others. The success of sustainable machining methods can be achieved if all components of machining or manufacturing system are suitable for such technologies.

For this, the whole processes and all aspects and elements involved have to be considered. First of all, a machine tool (manufacturing cell, production line etc.) and cutting tools should be specifically designed. Although some machine tools have been retrofitted, this solution does not appear to be an attractive alternative. Ideally, the consideration of ecological aspects in the area of machining starts from the part design. This should make the process easier in terms of chip removal (evacuation) and process performance.

Studies in this area are focused on process level activities and improvements. These improvements, including minimizing the use of cutting fluids (coolants or lubricants), optimizing material use, appropriate treatment of contaminated parts and chips and reducing cutting energy, do have important environmental links.

For example, metal working fluids have become more problematic in terms of both workers health and environmental pollution, with serious issues resulting from their use, treatment and disposal, are often analyzed as a problem for potential improvement. Beside ecological aspects, due to the high disposal costs, manufacturers are self-motivated to reuse products, reduce and/or eliminate waste, but they need alternative technologies from which to choose the best solution. Such processes and facilities must minimize flows and environmental loads wherever possible.

The overall sustainability level of the machining process is to be evaluated. The existing and new proposed process sustainability assessment methods should be used in order to involve science-based sustainability principles, not only for product manufacture but also for its design. In this respect it reveals the need for identifying relevant sustainability indicators and for developing methodologies for quantification of influence factors using modeling and optimization methods.

Abbreviations

CLF	Cooling lubrication fluids
CM	Cryogenic machining
DM	Dry machining
HPJAM	High pressure jet assisted machining
HSS	High speed steel
LN	Liquid nitrogen
NDM	Near dry machining
MQL	Minimal quantity lubrication
MQC	Minimal quantity cooling

References

1. Gutowski TG (2004) Design and manufacturing for the environment. In: Handbook of mechanical engineering. Springer-Verlag, Berlin
2. Jawahir IS, Wanigarathne PC, Wang X (2006) Product design and manufacturing processes for sustainability, mechanical engineers'. In: Kutz M (ed) Handbook: manufacturing and management, third edn, vol 3, John Wiley & Sons, Inc.,Delmar, pp 414–439
3. Dahmal P, Escursell M (2004) High pressure jet-assisted cooling: a new possibility for near net shape turning of decarburized steel. Int J Mach Tools Manuf 44(1):109–115
4. Dahmus IB, Gutowski TG (2004) An environmental analysis of machining. In: Proceedings of the 2004 ASME IMECE, Anaheim, pp 1–10
5. Graham I, Dom K (2000) Going dry. Manuf Eng 124(1):72–78
6. Jawahir IS, Dillon Jr OW (2007) Sustainable manufacturing processes: new challenges for developing predictive models and optimization techniques. In: Proceedings of 1st international conference on sustainable manufacturing SM1, Montreal, pp 1–15
7. Kundrák J, Mamalis AG, Gyáni K, Markopoulos A (2006) Environmentally friendly precision machining. Mater Manuf Processes 21(1):29–37
8. Pusavec F, Kramar D, Krajnik P, Kopac J (2010) Transitioning to sustainable production—part II: evaluation of sustainable machining technologies. J Clean Prod 18(12):1211–1221

9. Rajemi MF, Mativenga PT, Aramcharoen A (2010) Sustainable machining: selection of optimum turning conditions based on minimum energy considerations. J Clean Prod 18(11–12):1059–1065
10. Kotaiah KR, Srinivas J, Babu KJ, Srinivas K (2010) Prediction of optimal cutting states during inward turning: an experimental approach. Mater Manuf Processes 25(6):432–441
11. Snig KW, Rummen S (1994) Ecological manufacturing. Production Engineering. German Academic Society Engineering, Berlin, vol 1(2), pp 1–4
12. American Automobile Manufacturers Association (1995) Industrial metalworking environment: Assessment and control. Dearborn, USA
13. Byme G, Scholta E (1993) Environmentally clean machining processes—a strategic approach. Annals of CIRP 42(1):471–474
14. Pfeifer T, Eversheim W, Kanig W, Weck W (1994) Manufacturing excellence: the competitive edge. Chapman and Hall, London
15. Grzesik W (2008) Dry and semi-dry machining, advanced machining processes of metallic materials (2008), pp 226–245
16. Aronson RB (1995) Why dry machining? Manuf Eng 114(1):33–36
17. Kalhofer E (1997) Dry machining principles and applications. In: Proceedings of the 2nd Seminario International de Alta Tecnologia UNIMEP. Julho, Santa Barbara D'Oeste
18. Davim JP, Sreejith PS, Silva J (2007) Turning of brasses using minimum quantity of lubricant (MQL) and flooded lubricant conditions. Mater Manuf Processes 22(1):45–50
19. Pusavec F, Kopac J (2009) Achieving and implementation of sustainability principles in machining processes. J Adv Prod Eng Manag 3:151–160
20. Ahmad-Yazid A, Taha Z, Almanar IP (2010) A review of cryogenic cooling in high speed machining (HSM) of mold and die steels. Sci Res Essays 5(5):412–427
21. Tan XC, Liu F (2002) A decision-making framework model of cutting fluids selection for green manufacturing. J Mater Process Technol 129:467–470
22. Weinert K, Inasaki I, Sutherland JW, Wakabayashi T (2004) Dry machining and minimum quantity lubrication. CIRP Ann Manuf Technol 53(2):511–537
23. Macado AR, Wallbank J (1997) The effect of extremely low lubricant volumes in machining. Wear 210:76–82
24. Weinert K (1999) Trockenbearbeitung und Minimalmengekuehlschmierung—Einsatz in der spanenden Fertigungstechnik. Springerverlag, Berlin
25. Hong SY (2001) Economical and ecological cryogenic machining. J Manuf Sci Technol 123:331–338
26. Su Y, He N, Li L, Iqbal A et al (2007) Refrigerated cooling air cutting of difficult-to-cut materials. Int J Mach Tool Manuf 47:927–933
27. Pusavec F, Krajnik P, Kopac J (2010) Transition to sustainable production—part I: application on machining technologies. J Clean Prod 18(2):174–184
28. Courbon C, Kramar D, Krajnik P, Pusavec K, Rech J, Kopac J (2009) Investigation of machining performance in high-pressure jet assisted turning of Inconel718: an experimental study. Int J Mach Tools Manuf 49:1114–1125
29. Kaebernick H, Kara S, Sun M (2003) Sustainable product development and manufacturing by considering environmental requirements. Robot Comput Integr Manuf 19(6):461–468
30. Byrne G, Dornfeld D, Denkena B (2003) Advanced cutting technology. Ann CIRP 52(2):483–507
31. Byrne G, Scholta E (1993) Environmentally clean machining processes—a strategic approach. Ann CIRP 42(1):471–474
32. Kalpakjian S, Schmid S (2001) Manufacturing engineering and technology, 4th edn, Prentice Hall, Upper Saddle River
33. Chapman PF, Roberts F (1993) Metal resources and energy. Butterworth and Co., Ltd, Thetford
34. Environmental Protection Agency (EPA) (1995) Profile of the iron and steel industry. EPA Office of Compliance Sector Notebook Project, EPA/310-R-95-005, Washington

35. Becket GJP (1994) Corrosion. In: Byers JP (ed) Causes and cures from metalworking fluids. Marcel Dekker Inc., New York
36. Rossmore LA, Rossmore HW (1994) Metalworking fluid microbiology. In: Byers JP (ed) Metalworking fluids. Marcel Dekker Inc., New York
37. Childers JC (1994) The chemistry of metalworking fluids. In: Byers JP (ed) Metalworking fluids. Marcel Dekker Inc., New York, pp 165–189
38. Sherman J, Bain C, Huibers PDT, Garcia-Valls R, Hatton TA (1998) Solvent replacement for green processing. Environ Health Perspect 106(1):253–271
39. Steen B (1997) On uncertainty and sensitivity of LCA-based priority setting. J Clean Prod 5(4):255–262
40. Sluhan W (1997) Trample tramp oil. Cut Tool Eng 49(6):87–90
41. Klocke F, Eisenblatter G (1997) Dry cutting. CIRP Ann Manuf Technol 46(2):519–527
42. Oberg E, Franklin DJ, Holbrook LH, Ryffel H (1996) Machinery's handbook. In: Green RE (ed) 25th edn, Industrial Press Inc., New York
43. Stals LM, Nesladek M, Quaeyhaegens C (1997) Current industrial practice critical issues in hard PVD and PA-CVD coatings. Surf Coat Technol 9:230–239
44. Lahres M, Jorgensen G (1997) Properties and dry cutting performance of diamond-coated tools. Surf Coat Technol 96:198–204
45. Kalpakjian S (1995) Manufacturing engineering and technology, 3rd edn. Addison-Wesley Publishing Company, Reading
46. Frischknecht R, Rebitzer G (2005) The ecoinvent database system: a comprehensive web-based LCA database. J Clean Prod 13(13–14):1337–1343
47. D'Amico E (1995) Alternatives finding way in a stormy market. Chem Week 157(17):53–60
48. Kirschner EM (1994) Environment, health concerns force shift in use of organic solvents. Chem Eng News 72(25):13–20

Environmentally Friendly Machining: Vegetable Based Cutting Fluids

Emel Kuram, Babur Ozcelik and Erhan Demirbas

Abstract A wide variety of cutting fluids are commercially available in the cutting fluid suppliers in order to provide machining performances for a number of industries. In machining, mineral, synthetic and semi-synthetic cutting fluids are widely used but, recently, uses of vegetable based cutting fluids have been increased. Although, these cutting fluids are beneficial in the industries, their uses are being questioned nowadays as regards to health and environmental issues. Cutting fluids are contaminated with metal particles and degradation products which diminish the effectiveness of cutting fluids. To minimize the adverse environmental effects associated with the use of cutting fluids, the hazardous components from their formulations have to be eliminated or reduced to the acceptable level. In addition, mineral based cutting fluids are going to be replaced with vegetable based cutting fluids since they are environmentally friendly. Today to diminish the negative effects associated with cutting fluids, researchers have developed new bio based cutting fluids from various vegetable oils. This chapter has also focused on environmental conscious machining such as dry cutting, machining with minimum quantity lubricant and especially machining with vegetable based cutting fluids including other types of cutting fluids. Literatures associated with types of cutting fluids have also been presented in this chapter.

1 Introduction

The use of lubricants in metal cutting operations is relatively recent as compared to the use of fats to grease chariot wheels to date back to the times of the ancient Egyptians. The widespread use of cutting fluids coincided with the industrial revolution in the late Eighteenth century. Mineral oils were being extensively used as

E. Kuram · B. Ozcelik (✉)
Department of Mechanical Engineering, Gebze Institute of Technology,
41400 Gebze-Kocaeli, Turkey
e-mail: ozcelik@gyte.edu.tr

E. Demirbas
Department of Chemistry, Gebze Institute of Technology, 41400 Gebze-Kocaeli, Turkey

J. P. Davim (ed.), *Green Manufacturing Processes and Systems*, Materials Forming, Machining and Tribology, DOI: 10.1007/978-3-642-33792-5_2,

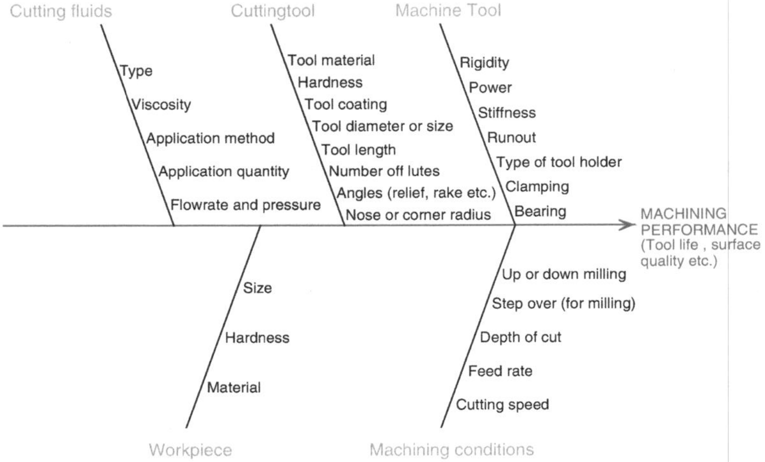

Fig. 1 Effect of machining parameters in performance

cutting fluids in the machining area by the mid-nineteenth century. In 1868, W.H. Northcott observed that the use of cutting fluids improved tool life. In 1883, Taylor used water in machining and demonstrated the importance of water as a cutting fluid and observed that cutting speeds could be increased by 30–40 % by using water.

The main aim of all machining operations is to obtain to lower machining costs by improving of quality and productivity. This aim can be achieved by machining at the highest cutting speed with long tool life, fewest part rejects (scrap) and minimum downtime. In machining, a lot of parameters affecting the cutting performances are shown in a fishbone (cause and effect) diagram (Fig. 1). Some machining operations can be carried out "dry", but cutting fluids have been used extensively and play a significant role in machining areas. Cutting fluids affect the productivity of machining operations, tool life, quality of workpiece and prevent the cutting tool and machine from overheating as well. The proper application of cutting fluid provides higher cutting speeds and higher feed rates possible. In general, a successful cutting fluid must not only improve the machining process performance, but also fulfil a number of requirements which are non-toxic, non-harmful to health for operators, not a fire hazard, not smoke or fog in use and cost less. One of the drawbacks of using cutting fluids is the waste disposal after being used.

Mineral based cutting fluids are reasonably priced so they are used extensively in machining area. But human beings were faced with mineral oil which was a limited resource due to the oil crisis of 1979 and 1983. Mineral oil has also poor biodegradability thus induces the potential for long term pollution of the environment. Moreover, the availability of mineral oil is highly dependent on political considerations. Therefore, existing deposits do not guarantee to us for the availability of mineral oil in the future [1]. The demand for biodegradable cutting fluids has increased with the use of vegetable based cutting fluids as an alternative to mineral based cutting fluids. Mineral based oils are limited and steadily decreasing resource whereas the vegetable based oils are sustainable.

2 Cutting Fluids

Classifications of cutting fluids are essential to understand them better since today a variety of cutting fluids are widely available in the world. According to chemical formulations, cutting fluids are classified into four categories: cutting oils, soluble oils (emulsified oils, emulsions), synthetic (chemical) fluids, semi-synthetic (semi-chemical) fluids. Cutting oils named as neat oil or straight cutting oil are formed oil derived from petroleum, animal or vegetable origin. Cutting oils used without further dilution in metal cutting processes have good lubrication properties, poor cooling properties and increases fire risk. They may also create a mist or smoke harmful to the health of operator. The use of cutting oils is limited to low temperature and low speed cutting operations.

Emulsified oils are a suspension of oil droplets in water. This cutting fluid is done by blending oil with emulsifier agent(s) to improve the stability of the emulsion in water. The general compositions of water based cutting fluids are as follows:

$$Base\ oil + Emulsifier + Other\ additives$$

$$Base\ oil \begin{cases} mineral\ oil \\ vegetable\ oil \end{cases}$$

$$Other\ additives \begin{cases} Neutralization\ agents \\ Corrosion\ and\ rust\ inhibitors \\ Lubricating\ additives\ (antiwear\ and\ EP\ additives) \\ Biocidies\ and\ fungicides \\ Foam\ inhibitors \end{cases}$$

Emulsifiers have the function of dispersing the oil in water in order to make a stable oil-in-water emulsion. Rao and Srikant [2] stated that thermal conductivity, kinematic viscosity and pH increased with an increase in the content of emulsifier whereas flash and fire points decreased with an increase in the amount of emulsifier. In turning of AISI 1040 steel decrements of cutting forces, surface roughness and tool wear with an increase in the emulsifier content were found by Srikant et al. [3]. Higher heat transfer rates, higher hardness and lesser surface roughness were observed in cutting fluids with higher rate of emulsifier content [4].

The concentrate cutting fluids must be stable without separating for a minimum of six months storage and emulsion stability is the most critical property of soluble oils. The presence of water in emulsions induces rust, bacterial growth and evaporation losses. Sulphur, chlorine and phosphorous based chemical additives known as extreme pressure (EP) additives are used under extreme pressure conditions. EP additives form solid lubricant layer between cutting fluid and the metal surface by chemical reaction. This film possesses low shear strength and good antiweld properties so EP additives can reduce friction and wear effectively. Emulsions have some advantages:

- reduction of heat allows higher cutting speeds in machining.
- dilution with water to cut the cost.
- no fire hazard and a lower rate of oil misting.

One of the drawbacks associated with emulsions is the fungi and bacteria growth which increases health hazards and diminishes the service life of cutting fluids. Presence of bacteria in the cutting fluid can cause separation in the emulsions. As a result the coolant lubricity capability is degraded by the bacteria. Moreover, pH of the coolant can help to reduce corrosion of workpiece and machine tool, and influences the microbial activity [5]. Germicide and bactericide additives are added to emulsions to control the bacteria growth.

In order to control bacterial growth in cutting fluids, chemical additives are necessary but they are hazardous for both the environment and health of operators [5]. Antimicrobials and biocides are utilized to maintain the efficiency of cutting fluids rather than protecting the operators. Formaldehyde releasing biocides are potential carcinogenic. Some of lubricants in the cutting fluids is considered to be hazardous to the environment and health [6] such as chlorinated extreme pressure additives.

Synthetic and semi-synthetic cutting fluids are mentioned in Sect. 2.2. Advantages and disadvantages of different types of cutting fluids are presented in Table 1.

Cutting fluids are applied to the cutting region in order to improve the cutting performance. The primary function of cutting fluid is to reduce temperature generated at cutting tool/workpiece interface. The hardness and resistance to abrasion of cutting tools are reduced at high temperature. Temperatures generated during machining affect the tool wear. So the reduction of this temperature will cause extending tool life. Cutting fluids also cool the workpiece, thus preventing its final dimensions. Cutting fluids' cooling of workpiece function is very important especially in grinding operations. The reduction temperatures ability of a cutting fluid during machining depends on its thermal properties especially specific heat and thermal conductivity. The other function of cutting fluid is lubrication. Lubrication effect of cutting fluids minimizes the amount of heat generated by friction. Cutting fluids with high lubricant ability are generally used in low-speed machining such as screw cutting, broaching and gear cutting and on difficult-to-cut materials, whereas cutting fluids with high cooling ability are generally used in high-speed machining [7].

Table 1 Advantages and disadvantages of cutting fluids

Straight oils	Soluble oils	Semi-synthetics	Synthetics
Advantages			
Excellent lubricity	Good lubricity	Good cooling	Excellent cooling
Excellent rust control	Good cooling	Good rust control	Excellent microbial control
		Good microbial control	Nonflammable, nonsmoking
			Good corrosion control
			Reduced misting and foaming problems
Disadvantages			
Low cooling	Rust control problems	Foam easily	Poor lubricity
Fire hazard	Bacterial growth	Stability is affected by water hardness	Easily contaminated by other machine fluids
Create a mist or smoke	Evaporation losses	Easily contaminated by other machine fluids	
Limited to low-speed and heavy cutting operations			

Cooling effect can be done best by water with low cost but its lubrication properties are very low. Water possesses high specific heat and thermal conductivity and this is the reason why water is used as the base in cutting fluids. Besides, water is cheap, supplied easily and its low viscosity provides it to flow at high rates. However, it causes some corrosion at ferrous metals and this may be diminished with corrosion inhibitors.

Cutting fluids consist of base oil(s), emulsifiers, corrosion inhibitors, lubricating, anti-wear and high pressure additives, neutralizing agents, biocides, fungicides, foam inhibitors and stabilizing agents to obtain favourable properties and to diminish the harmful effects.

The chemical additives used to formulate cutting fluids provide various functions such as emulsification, corrosion inhibition, lubrication, microbial control, defoaming, dispersing and wetting. Most of the additives used are organic chemicals that are anionic or nonionic in charge and most of the additives are liquids in order to blend easy. Some of the chemical additive types used are fatty acids, esters, sulfonates, soaps, chlorinated paraffins and fatty oils.

Cutting fluids also remove chips from tool/workpiece interface to prevent a finished surface. Especially at higher cutting speeds and feed rates, greater amounts of chips are generated in machining. Hence removal of chips from cutting area at these situations is very significant function of cutting fluids. The cutting fluid capability of flushing away the chips from the cutting zone depends on mainly its viscosity and flow rate. Viscosity is the resistance to flow of oil and is affected inversely by temperature. The tendency of the viscosity of oil with temperature changes is called *viscosity index* (V.I). A low V.I. signifies a relatively large change of viscosity with changes of temperature. In other words, the oil becomes extremely thin at high temperatures and extremely thick at low temperatures. A high V.I. means relatively little change in viscosity over a wide temperature range.

Cutting fluids reduce the adhesion of the workpiece to the cutting tool (reduce to form the built up edge (BUE) on the cutting tool), protect the workpiece and cutting tool surfaces from corrosion and lower the power required to machine (decreases friction). This shows both energy saving and less heat generating. If less heat is generated during machining, tool life of the cutting tools is longer and surface integrity of the workpiece is improved. Overall, the machining process will tend to be more stable.

Cutting fluids should have the following properties to fulfill their functions properly:

- Good lubricating properties
- High cooling capacity
- Low viscosity to provide free flow of cutting fluid
- Chemically stable
- Non-corrosive
- High flash point to reduce fire risks
- Allergy free
- Less evaporative
- Low cost

The selection of appropriate cutting fluid is very important because it could affect machining performance (tool life, cutting forces, surface roughness, power consumption etc.) and the selection depends on some parameters such as workpiece material used, cutting tool material and type of machining process. For instance, cutting fluids containing sulfur and chlorine additives should not be used with nickel-based alloys and titanium, respectively. Cutting fluids with high lubricity ability are generally used in low-speed machining such as screw cutting, broaching and gear cutting and on difficult-to-cut materials, whereas cutting fluids with high cooling ability are generally used in high-speed machining.

2.1 Mineral Based Cutting Fluids

Mineral based cutting fluids consist of oils extracted from petroleum. Mineral oils are hydrocarbons and their properties base on the chain length, structure and refining level. Two types of mineral oil are used in metal cutting fluids: Paraffinic and naphthenic. Paraffinic oils consist of long linear chains of hydrogen and carbon atoms. Naphthenic oils behave differently from paraffinic oils due to the molecular structure with hydrocarbons rings. Mineral oil has poor biodegradability thus it induces the potential for long term pollution of the environment. Mineral based oil is also a limited and steadily decreasing resource.

2.2 Semi-Synthetic and Synthetic Cutting Fluids

Synthetic and semi-synthetic cutting fluids are blended with water and various chemical agents. These agents are added for rust prevention, lubrication and reduction of surface tension. Synthetic cutting fluids have good coolant properties but their lubricant properties are less than the other cutting fluid types. Since synthetic cutting fluids are transparent, they help the operator to monitor the machining process. Synthetic cutting fluids are generally more resistant to biological attack than emulsions. Semi-synthetic cutting fluids are combinations of synthetic cutting fluids and emulsions. Semi-synthetic cutting fluids contain less oil (2–30 % oil) whereas synthetic cutting fluids contain no oil.

2.3 Vegetable Based Cutting Fluids

Vegetable oils consist of triacylglycerides (triglycerides) which are glycerol molecules with three long chain fatty acids attached at the hydroxyl groups via ester linkages. The fatty acids in vegetable oil triglycerides are all of similar length, between 14 and 22 carbons long. But their unsaturation levels vary. The triglyceride structure of vegetable oils provides desirable properties of lubricant. Long, polar fatty acid chains provide high strength lubricant films which interact strongly with metallic surfaces and reduces both friction and wear [8]. Vegetable oils have a higher viscosity index. However, thermal and oxidation stability of vegetable oils are limited [9]. Vegetable oils perform better than the other oils and the reasons are described as follow:

- Vegetable oils have good lubricity properties. The highly lubricating properties of vegetable oil are made possible by the fundamental composition of the vegetable oil molecules, as well as the chemical structure of the oil itself. Its properties are the direct result of the vegetable oil's smart molecules. These molecules are long, heavy, and dipolar in nature; that is, the ends of the molecules have opposing electrical charges [10]. Vegetable oils carry slight polar charge but mineral oils have no charge. This polar charge draws the vegetable oil molecule to a metallic surface like little magnets; therefore, vegetable oils adhere to a metal surface more tightly than mineral oils [10, 11]. Dense, homogeneous alignment of vegetable oil molecules creates a thick, strong and durable film layer of lubricant. This lubricating film gives the vegetable oil a greater

capacity to absorb pressure. In contrast, the molecules of mineral oils are intrinsically non-polar. They form a random alignment along a metal surface, which provides a weaker layer of lubrication [10]. Consequently, vegetable oils make a better lubricant [11].
- Vegetable oils have a higher flash point, which reduce smoke formation and fire hazard [10, 11]. Higher flash point value allows using the cutting fluid in high temperature conditions.
- Viscosity is another oil property that has an important effect on machining productivity [10]. Vegetable oils have a high natural viscosity as the machining temperature increases. The viscosity of vegetable oils drops more slowly than that of mineral oils. As the temperature falls, vegetable oils remain more fluid than mineral oils, facilitating quicker drainage from chips and workpieces. The higher viscosity index of vegetable oils ensures that vegetable oils will provide more stable lubricity across the operating temperature range [11].
- Vegetable oil molecules are quite homogeneous in size but mineral oil molecules vary in size. Consequently, the properties of mineral oil such as viscosity, boiling temperature are more susceptible to variation [10].
- Vegetable oil has higher boiling point and greater molecular weight and this results in less loss from vaporization and misting [12].

3 Environmental Aspects of Cutting Fluids

The use of cutting fluids repeatedly over time induces chemical changes of cutting fluids. These changes are due to the environmental effects, contamination from metal chips and tramp oil. The growth of bacteria and yeast becomes environmental hazard and also adversely affects the effectiveness of the cutting fluids. Cutting fluids degrade in quality with use and time and when they lose their quality the disposal of them is mandatory. Waste disposal of cutting fluids are expensive and affect the environment negatively.

The focus on lubricants has shifted from biodegradability to renewability over the years and owing to the change in human beings' environmental thinking [13]. Several aspects of an environmentally adapted lubricant are as follows [14, 15]:

- Biodegradability
- Toxicity
- Renewability
- Bioaccumulability and biomagnifications
- Life cycle assessment (LCA)
- Energy saving and fuel economy

Biodegradability is the degradation by the action of micro organisms [15]. Environmental compatibility of cutting fluids is determined mostly with biodegradability [16]. Alves and Oliveira [17] conducted a biodegradation test in dark

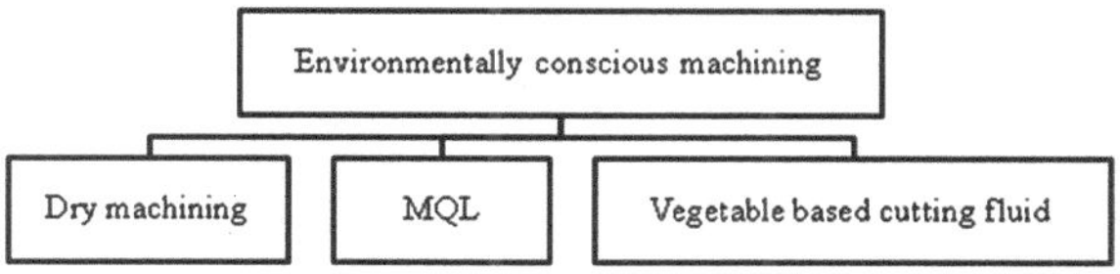

Fig. 2 Environmentally conscious machining

at 20–25 °C for 28 days for a new cutting fluid from castor oil. This cutting fluid showed high degradation rates and under these conditions mineral oil were degraded to 20–60 % hence mineral oil was not regarded as readily biodegradable. In an another work, it was found that vegetable based synthetic ester and rapeseed oil had 100 % biodegradable, whereas neat type of cutting oil had 20–30 % biodegradable [18]. Bioaccumulation is a substance accumulation in an organism. Biomagnification is to increase the concentration of accumulated substance in the food chain [14]. Renewability is the relative amount in any given product of raw material that can be re-grown, recycled or re-used [15].

Cutting fluids affect the health of operator negatively in machining operations which can be vaporised, atomised and form mist owing to high pressure and temperature. This airborne particle of cutting fluids can be inhaled by operators and causes mild respiratory problems, asthma and several types of cancers (oesophagus, stomach, pancreas, colon, etc.) [6]. Mist, fumes, smoke and odors can cause severe skin reactions and respiratory problems. When physical contact with cutting fluid occurs, dermatological problems are seen in operators. Cutting fluids also may influence adversely the machine tool components which should be cleaned to remove any cutting fluid residue. This cleaning operation requires additional time and cost. Water based and low viscosity cutting fluids can be preferred in order to ease of cleaning.

Cutting fluids used in machining area contain environmentally harmful chemical substances. These chemicals have negative effect on the environment and human health as well. Most of the cutting fluids used in machining are petroleum origin and the disposal of petroleum-based cutting fluids causes water contamination, air and soil pollutions. Dry cutting, machining with minimum quantity lubricant (MQL) and vegetable based cutting fluids are believed as environmental conscious machining (Fig. 2).

4 Application Methods for Cutting Fluids

Cutting fluids can be applied to a cutting tool/workpiece interface by some methods such as manual, flooding and mist application. For efficient machining performance, a lubricant film must be formed at the sliding surface. For better performance, application of cutting fluid to the cutting zone must be continuous, not intermittent. Unless cutting fluid is carefully placed, it cannot perform its cooling and lubrication functions effectively.

In manual application, operator uses oil container to apply cutting fluid to the cutting region. This is the easiest and the cheapest method of cutting fluid application; however it has limited use in machining area. Manual method is intermittent cutting fluid application so its performance is low compared to continuous application methods. Access of cutting fluid to the cutting region is limited in this application.

Flooding is the most common application method of cutting fluids to the tool/workpiece interface. In flooding application, large quantity of cutting fluid is continuously delivered to the cutting region by means of a pipe, hose or nozzle. The cutting fluid is cumulated in a reservoir, filtered and pumped back to the delivery nozzle in this method. In order to obtain optimum machining performance, direction of nozzle, number of nozzles and flow rate of cutting fluids must be optimized. At some situation cutting fluid is not able to reach cutting zone effectively.

In mist application, cutting fluids are atomized and blown onto cutting tool/workpiece interface. This method is not as effective as flooding to cool the cutting tool; however it may sometimes be more effective than flooding such as delivering the cutting fluids to cutting zone that are difficult to access by flooding. However, inhalation of mist by the operator induces health problems so very efficient ventilation is required.

Cutting fluids can be applied to cutting region three possible directions as shown in Fig. 3.

- on the back side of the chip,
- along between the chip and rake face of the cutting tool,
- along between the finished workpiece surface and flank face of the cutting tool [19].

Mendes et al. [20] investigated the performance of cutting fluids in the drilling of AA 1050-O aluminium and applied cutting fluid as a mist. They also investigated

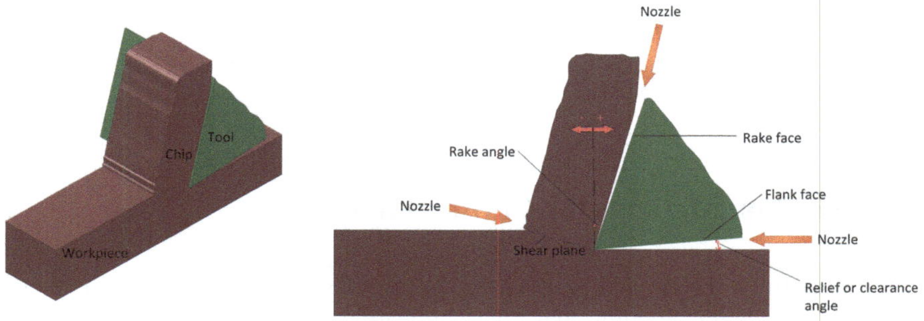

Fig. 3 Application directions of cutting fluids in orthogonal cutting

the effect of additives (chlorine, sulphur and phosphor) on the performance of the cutting fluid applied as a flood in the turning of 6262-T6 aluminium alloy. In drilling, 100 ml/h flow rate resulted in lower feed forces especially at higher cutting speeds and feed rates. In contrast to the feed force results, increment in the cutting flow rate in general resulted in higher torque, power consumption and specific cutting pressure. Surface roughness was not significantly affected by cutting fluid flow rate in the drilling. In general, an increment of the cutting fluid concentration showed a decrement of the cutting force, but this decrement was nearly negligible when comparing concentrations of 10 and 15 % in the turning. Three directions of cutting fluid application were compared: (1) over the chip and rake face, (2) at the tool-chip interface and (3) at the tool-workpiece interface. When cutting fluid was applied over the chip and rake face, considerably higher forces were observed. Experimental turning study was also carried out in which the cutting fluid concentration (10 %) was kept constant as well as the direction of cutting fluid application (at the tool-workpiece interface) so that compared the effect of EP additives (chlorine, sulphur and phosphor). The lowest cutting force was achieved using the cutting fluid with chlorine additive followed by the cutting fluid with sulphur additive. The best surface finish was obtained using the cutting fluid with chlorine additive.

Especially in drilling the use of cutting fluids under high pressures (pressurized jets) via internal holes in drills can improve lubrication, cooling and chip removal. The cutting fluid is fed through internal holes to the cutting region. In high-pressure cutting fluids, greater penetration of the cutting fluid into the tool-chip interface occurs as compared to the flooding application. Machado et al. [21] found that using high pressure cooling when turning of Ti6Al4V increased tool life up to 300 % as compared to conventional flood application and lower tool life was achieved when machining Inconel 901 under the high pressure coolant jet. Kaminski and Alvelid [22] proved that when applying high-pressure coolant at 250 bar, the cutting temperature could be reduced by ~40 % compared to flood application and further increment in pressure had minimal additional effect. Ezugwu and Bonney [23] found that acceptable surface finish and improved tool life can be obtained during machining of Inconel 718 with high coolant pressures. When machining at 203 bar coolant pressure at a cutting speed of 50 m/min, tool life increased 740 % as compared to conventional coolant application. In general, increment in tool life was observed with increasing coolant pressure. In another study, Ezugwu and Bonney [24] carried out the turning experiments of Inconel 718 alloy under conventional and high-pressure (11, 15 and 20.3 MPa) coolant supplies. Tool life, surface roughness and force components were measured and it was concluded that acceptable surface finish and improved tool life could be achieved when machining with high coolant pressures. The highest improvement in tool life (349 %) was obtained when turning with 11 MPa coolant pressure at higher cutting speed of 60 m/min. Machining with coolant pressures in excess of 11 MPa at cutting speeds up to 40 m/min decreased tool life more than machining with conventional coolant flow. This result showed that there was a critical coolant pressure which the cutting tools performed better under high-pressure coolant supplies.

However, in milling of En32b low carbon steel, flood coolant gave lower flank wear than high-pressure (1.8 MPa) environment [25]. Effect of fluid pressure, flow rate and direction of application in finish turning of AISI 1045 steel were investigated. When cutting fluid was applied to the tool rake face, the adhesion between chip and tool was very strong, causing the removal of tool particles and large crater wear. When cutting fluid was not applied to the rake face, adhesion of chip material to the face occurred, but was not strong enough to remove tool particles as it moved across the face, and thus crater wear did not increase [26]. The effects of ultra-high pressure coolant on the surface integrity and tool life were investigated during finish turning of Inconel 718. Conventional flood cooling and ultra-high pressure coolant were supplied at 5 bar and from 70 to 450 bar, respectively. Also the effects of applying ultra-high pressure coolant to the rake face alone, flank face alone and both positions together were investigated. Applying ultra-high pressure coolant to the rake face of the tool (rake only) decreased the tool life with an increment in pressure; however this was not observed when the cutting fluid was applied at the flank face of the tool (flank only) or at both the flank and rake faces of the tool simultaneously (flank and rake). When the high-pressure jet was applied to the flank face, flank wear decreased but the notching was not affected. When using up to 450 bar pressure, no increment in tool life was seen. The results also showed that the level of workpiece microstructural deformation or surface roughness obtained when machining with either new or worn tools were not affected beneficial or detrimental by the application of ultra-high pressure coolant. Cutting fluid pressure and direction had relatively little effect on the level of surface integrity [27]. Machining experiments were conducted under conventional wet, high-pressure neat oil and high-pressure water soluble oil environments during turning of Ti-6Al-4V. High-pressure neat oil environment provided longest tool life [28].

5 Minimising Adverse Environmental Effects of Cutting Fluids

Environmental concerns, market forces and legislative requirements make imperative a search for new solutions that minimize environmental impact [29]. To minimise the adverse environmental effects associated with the use of cutting fluids, the best solution is to remove the hazardous components from their formulation. But we focused on alternative methods rather than formulation. The effective way of minimising adverse environmental effects from the use of cutting fluids is to minimise the volumes used and replace mineral based cutting fluids with environmentally friendly cutting fluids such as vegetable based cutting fluids. From an environmental point of view, the best method is dry machining. Dry machining not only reduces the contamination of water and air but also reduces danger to health of the operator. However, dry machining is not efficient for many cutting operations. In this case, minimum quantity lubricant (MQL) can be penetrated into the cutting zone so that improve machinability. MQL reduces the cutting fluid consumption but this method uses cutting fluids in the form of mist which increases health hazards for the operators [30]. Waste treatment costs, negative environmental effects and health hazards of petroleum based cutting fluids increase the requirement for renewable and biodegradable lubricants and vegetable based cutting fluids have a higher potential of use under these limitations [31]. Vegetable based cutting fluids can be considered environmentally friendly since these fluids are renewable and have high levels of biodegradability.

5.1 Dry Machining

Dry machining means that no cutting fluid is used during process. For economic as well as environmental reasons machining process is carried out without any cutting fluid but dry machining has some disadvantages. During dry machining process, temperature of the cutting tool is very high and this induces excessive tool wear thus decreasing tool life. Also the chips generated at machining cannot wash away and these chips cause deterioration on the machined surface.

The problems of cutting fluid contamination and disposal are not seen in dry machining. Dry machining does not induce the pollution of atmosphere or water resources. Contrary to dry machining in wet machining (machining with cutting fluids by any means flooding and MQL), environment, water source and soil become polluted during disposal of the cutting fluid. Application of machining with dry will also diminish the manufacturing costs.

In some cases for instance in interrupted machining process like milling, dry cutting gives longer tool life than machining with cutting fluid. In milling cutting tool does not cut continuously and the using of cutting fluids increase thermal shock effect. Hence, dry machining is better suited for milling operations. In addition machining with ceramic tools must be conducted in dry condition due to the thermal shock. In drilling especially gun drilling the most important function of cutting fluid is the chip removal and dry cutting may induce drill breakage.

Dry cutting shows positive effects in some workpiece materials such as AISI 316 stainless steel [32]. Diniz et al. [33] used two concentrations (7 and 12 %) of the vegetable oil based emulsion with two different ways of fluid application (internally and externally to the tool) in the milling of 15-5PH stainless steel and the tool life results were compared to the dry cutting. Tool life for dry cutting was 3.5 times higher than those obtained when abundant fluid was utilized. It was found that the way of fluid application did not influence tool life.

However, in some workpiece materials dry machining presents many problems. For instance, aluminium is a soft material and dry machining of aluminium induces BUE. This influences on the surface quality of the workpiece.

Higher friction between tool and workpiece in dry machining can increase the temperature in cutting region. This high temperature will cause dimensional inaccuracies at the workpiece and excessive tool wear. So the disadvantages of dry machining have to be compensated. Improving cutting tools properties by better tool materials with lower friction coefficient and high heat resistance, coatings or tool geometries are investigated by researchers in order to compensate the effects of the elimination of cutting fluids in machining.

In order to make the application of dry machining feasible, some attempts have been done by the researchers. One of these attempts is to use tool materials with coating. In dry machining, advanced cutting tool materials and selection of cutting parameters are inevitable; however these cutting tools are very expensive and increase the machining costs [6].

Diniz and Micaroni [34] carried out turning experiments of AISI 1045 steel at varying cutting speed, feed and tool nose radius in order to obtain cutting conditions more suitable for dry cutting which make tool life closer to cutting fluid conditions without damaging the workpiece surface roughness and not increasing cutting power consumed by the process. To reach these aims it was mandatory to increase feed and tool nose radius and decrease cutting speed when cutting fluids are removed from a turning process. The use of cutting fluids gave longer tool life than dry cutting but difference in tool life between wet and dry cutting reduced at higher feed values. Dry cutting also showed less power and surface roughness than wet cutting. The increment of tool nose radius increased tool life and cutting power.

5.2 *Machining with Minimum Quantity of Lubrication*

In MQL method, a small amount of cutting fluid (10–100 ml/h) with compressed air is applied to the chip-tool interface to lubricate the contact area of chip-tool, to reduce temperature and friction. MQL can reduce the cost associated with the disposal of waste oils and the cutting fluid cost, whereas carrying chips away from cutting regions is limited.

In MQL; chip, workpiece and tool holder have a low residue of lubricant thus their cleaning is easier and cheaper as compared to flooding of cutting fluid. The cutting region is not flooded in MQL during machining so the operation can be seen by the operator [35]. MQL is used as a lubricating method rather than cooling. This poor cooling capacity limits the effectiveness of MQL in machining of difficult-to-machine materials such as titanium and nickel based alloys due to the excessive heat generation [36]. Thus in machining these pros and cons must be taken into consideration. The cutting performance of MQL mainly depends on nozzle pressure, number of pulses and amount of cutting fluid in each pulse so it is possible to produce high quality components with MQL by carefully choosing these parameters [37].

Since the negative effects associated with the cutting fluids, a lot of study has been concentrated on minimizing the use of cutting fluids or to eliminate cutting fluids in machining. Several experimental studies have investigated for the performance of MQL in the drilling [38–45], turning [46–51], milling [52–57] and grinding [58–61] processes. The most literature studies compared the performance of MQL with dry cutting and flood application. Conflicting results were found in the literature regarding the effect of different cutting fluid application methods (dry, MQL, flood) on cutting performance.

Kelly and Cotterell [38] used vegetable oil as MQL lubricant during drilling of cast aluminium silicon alloys. They concluded that the location of the feed nozzle, volume flow and pressure of the cutting fluid could be optimized in order to achieve longer tool life. Braga et al. [39] used MQL technique in the drilling of aluminium–silicon alloys. Zeilmann and Weingaertner [40] investigated the temperature during drilling of Ti6Al4V with MQL and it was found that internal MQL gave the lower temperature than external MQL. Davim et al. [29] reported experimental results of dry, MQL and flood-lubricated conditions during drilling of AA1050 aluminium. The cutting power and specific cutting force were higher for dry drilling but MQL and flood-lubricated conditions did not show much variation. MQL and flood-lubricated conditions gave similar surface finish results. A proper selection of the range of cutting parameters gave similar performances to flood-lubricated conditions by using MQL. Heinemann et al. [41] investigated the effects of MQL and dry cutting on the tool life in drilling of plain carbon steel. A discontinuous supply of the MQL showed a significant reduction in tool life as compared to a continuous supply of the MQL. It was also concluded that a low-viscous type with a high cooling capability type of MQL prolonged tool life. Tasdelen et al. [42] evaluated the effect of MQL, emulsion and air cooling on tool wear, surface finish and cutting forces during drilling of hardened steel. The lowest force and tool wear were obtained with emulsion and MQL, respectively. Costa et al. [43] studied the height of the burr under dry machining, MQL at the flow rate of 30 ml/h and conventional way (flooding) in the drilling. Vegetable oil in MQL, mineral oil in MQL and flooding and semi-synthetic oil in flooding were used as a cutting fluid. The smallest burr height was obtained for the dry drilling and the largest for the MQL systems. It was found in the literature that surface roughness, torque, force and tool wear of MQL drilling of austempered ductile iron were lower than that of dry drilling but higher than flooding [45].

Low quantity of cutting fluid is applied as a mist application in the most literature [38, 62, 63]. However, applying cutting fluid in the form of mist poses serious health hazards such as irritation, respiratory problems. Varadarajan et al. [46] introduced a new minimal cutting fluid application technique which diminished the problems caused by mist application during hard turning of hardened tool steel (AISI 4340). In this new method, a small quantity of cutting fluid was applied in the form of a high-velocity, narrow, pulsed jet. The rate of injection of cutting fluid, the injection and pulsing rate were 2 ml/min, 20 MPa and 600 pulse/min, respectively.

Dry and MQL conditions gave always smaller flank wear and surface roughness values than the wet cutting in the turning of SAE 52100 hardened steel [47]. Khan and Dhar [12] investigated the effect of MQL by vegetable oil on cutting temperature, tool wear, surface roughness and dimensional deviation in turning of AISI 1060 steel. MQL reduced the cutting temperature, tool wear and surface roughness as compared to dry machining. In turning of normalized 100Cr6 steel when MQL is applied to the tool rake, tool life is generally no different from dry cutting; however MQL applied to the tool flank can increase tool life [35]. Kamata and Obikawa [49] used dry, wet and MQL conditions during finish-turning of Inconel 718. The longest tool life was achieved with wet cutting but the surface finish was not good. It was also found that there is the optimum air pressure in finish-turning of Inconel 718 with MQL. Sreejith [50] analyzed the effect of dry machining, MQL and flooded coolant conditions with respect to the cutting forces, surface roughness and tool wear in turning of 6061 aluminium alloy. In MQL, the amount of material adhered was observed to be more compared with flooded and less compared with dry condition. It was not seen any considerable reduction in the adhered material, as the quantity of the lubricant was increased from 50 to 100 ml/h in MQL. The flank wear was seen to be almost same with MQL and flooded application. The lowest and highest resultant forces were achieved with flooded application and dry machining, respectively. For improving the quality of the workpiece surface, it was found that flooded coolant application was essential. Flood application of cutting fluid gave the lowest wear among the other application methods (MQL, MQL_EP and dry) during turning of AISI 1045 steel [51].

The efficiency of MQL was investigated in high speed milling of wrought aluminium alloys and was compared to emulsion. The effect of the position of the injection nozzle in relation to the feed direction was also studied. The nozzles were located at 45 and 135° in relation to the feed direction. In MQL, oil consumption was fixed at two values: 0.06 and 0.04 cm^3/min. Flank wear with MQL was always smaller than flank wear with the emulsion. The optimum nozzle-feed position was 135° considering tool life. Oil consumption below 0.06 cm^3/min showed a small increase of flank wear, but it was not very significant [52]. The amount of lubricant used in the study was much lower than that mentioned in most other studies [38, 46, 48, 64]. The study done by López de Lacalle et al. [52] not only improved tool life but also reduced the consumption of cutting fluids by 95 %.

Flank wear results showed that 6 ml/h was the best choice for the oil quantity among the others (6, 12 and 24 ml/h). P20 tool steels were milled with MQL at an oil volume of 6 ml/h and the distance between the nozzle and the tool tip was varied. When the nozzle was placed 60 mm away from the tool tip, the flank wear reached the maximum value. The flank wear was low and almost constant when the distance was between 80 and 200 mm. An appropriate oil quantity and distance between the nozzle and tool tip provide the optimum process condition in MQL [53]. The performance of minimal cutting fluid application in pulsed-jet form in the high-speed milling of hardened steel was investigated [55]. Flank wear in pulsed-jet application was lower than in flood application and dry cutting, especially at high cutting speed and/or low feed rate. The performance of pulsed-jet application was superior to that of flood application and dry cutting in terms of surface finish. Experiments were also conducted until tool failure so that investigate the progression of cutting force, surface roughness and flank wear against cutting time at constant machining parameters. Cutting forces in pulsed-jet application were lower than in flood application and dry cutting considering the whole period of machining. Increment of flank wear in pulsed-jet application was slow at the beginning and after 50 min machining flank wear increased at a higher rate. Whereas, increment of flank wear in flood application and dry cutting was rapid at the beginning and after around 30 min, rate of flank wear became slower. When tool life criterion was set at 0.35 mm of flank wear, tool life of all applications was found to be almost the same. Lowest surface roughness values were achieved with pulsed-jet application. They stated that the lubricity of cutting fluid had a dominant effect on tool wear rather than cooling. They also said that in flood application, cutting fluid may not be able to access the tool-chip interface owing to the low pressure of flood application. Sales et al. [56] evaluated tool wear, surface roughness and burr formation when milling AISI 4140 steel with cutting fluid applied by MQL technique. The vegetable based cutting fluid and different flow rate (dry cutting, 50, 100, 150 and 200 ml/h) were used. Increment in coolant flow rate tended to reduce tool wear, surface roughness and burr length.

MQL flank face cooling showed longer tool life and lower surface roughness as compared to dry and MQL rake face application since the cutting fluid could not reach the tool-chip interface during application of cutting fluid on the rake face [64].

The overall performance (cutting force, tool life, surface finish, cutting ratio, cutting temperature and tool-chip contact length) during MQL was found to be superior to dry and conventional wet turning of hardened steel [65]. Obikawa et al. [66] investigated the performance of MQL in high-speed grooving of carbon steel. It was found that tool wears reduced in MQL more effectively than the solution type cutting fluid. The tool wears decreased drastically with the increment the pressure of air supply.

Tawakoli et al. [58] investigated the MQL technique in grinding of a 100Cr6 hardened steel and a 42CrMo4 soft steel. Grinding forces and surface quality were measured under different environments (dry, MQL, fluid). The surface finish in grinding of 100Cr6 hardened steel was significantly better when MQL technique was used. However, in grinding of 42CrMo4 soft steel with MQL, the surface roughness was found to be higher than that in fluid application. MQL grinding gave lower tangential forces than both dry and fluid application. Hadad et al. [60] found in MQL higher temperature than that in fluid application during grinding of 100Cr6 (AISI 52100) steel. Mao et al. [61] investigated the grinding performance of AISI 52100 steel with respect to grinding force, temperature and surface integrity of workpiece under the different cooling-lubrication conditions (wet, dry, pure oil MQL and oil–water MQL). The lowest grinding force was obtained under wet condition, while dry grinding gave highest force. Pure oil MQL grinding had slightly lower tangential force than that of oil–water MQL. Surface roughness results were similar to force results. Wet grinding had the lowest temperature and dry grinding had the highest. Significant difference in temperature between pure oil MQL and oil–water MQL was not found in grinding.

5.3 Machining with Vegetable Based Cutting Fluids

Although attempts in manufacturing research are focused to diminish the use of cutting fluids, the present state-of-the-art technologies do not seem to assure that cutting fluids will be entirely eliminated in the next future [67]. Some machining operations and workpiece materials have still required the use of cutting fluid. New trend concerning the cutting fluids in machining is to replace hazardous components with environmentally friendly compounds. These new compounds not only must show the same properties to cutting fluids but also must improve the machining performance such as productivity.

Literature studies about machining using vegetable based cutting fluids are limited. Higher cost of vegetable based oils relative to mineral based oils is the principal limitation of them but this drawback will be diminished in the future as the petroleum prices increase [68].

Literatures associated with vegetable based cutting fluids in drilling, turning, milling and grinding have presented in this section.

5.3.1 Drilling

Kelly and Cotterell [38] used vegetable oil as MQL lubricant during drilling of cast aluminium silicon alloys. The effect of various methods of cutting fluid application (flood lubrication, MQL-mist, compressed air and dry) on cutting temperatures, torque, cutting forces and surface roughness were investigated. MQL using vegetable oil gave lower feed forces, torques and surface roughness at the higher cutting speeds and feed rates. Flooding with mineral oil showed lowest cutting temperature. Costa et al. [43] studied the height of the burr under dry machining, MQL at the flow rate of 30 ml/h and conventional way (flooding) in the drilling. Vegetable oil in MQL, mineral oil in MQL and flooding and semi-synthetic oil in flooding were used as a cutting fluid. The smallest burr height was obtained for the dry drilling and the largest for the MQL systems. The MQL with vegetable oil generally produced smaller burr heights than that of the MQL with mineral oil. Rahim and Sasahara [69] studied MQL palm oil (MQLPO) as a lubricant in the high speed drilling of Ti-6Al-4V and for the comparison purpose MQL synthetic ester (MQLSE), air blow and flood conditions were used. MQLPO gave lower tool wear rate than MQLSE and flood condition also showed low flank and corner wear rate. For flood condition, both wear rates laid between MQLPO and MQLSE however the tool life was the same. MOLPO exhibited lower tool wear rate than MQLSE and air blow conditions and comparable with flood condition. Significant improvement of the friction and wear in palm oil was due to the fatty acid content of palm oil. The carbon chain length of the fatty acids in palm oil is longer than the synthetic ester and this increment enhances durability of the contact. Reaction between metal oxide layer and the fatty acid leads smooth sliding and low friction. Metal soap has been formed on the contact surface owing to this reaction. Longer carbon chain can resist high cutting temperature so protects the surface. The molecular thin film present during the drilling under MQLPO reduced the friction and heat generation thus improved tool wear. Besides, the high viscosity of palm oil has a tendency to resist the flow, providing effective lubricating at the tool-chip interface, which reduces the friction, thus prevent the cutting tool from rapid wear. They found that MQL and flood condition have similar effects on the tool wear rate and tool life. The lowest thrust force and torque were obtained with flood condition. MQLPO exhibited comparable performance to the flood condition with respect to maximum workpiece temperature. Belluco and De Chiffre [67] determined the efficiency of vegetable based cutting fluids in drilling of AISI 316L austenitic stainless steel by measurement of tool life, tool wear, cutting forces and chip formation. A commercial mineral based cutting fluid was taken as a reference fluid. It was found that all vegetable based cutting fluids performed better than the reference mineral based cutting fluid. The best performance was achieved with a vegetable based cutting fluid giving 177 % increment in tool life and 7 % reduction in thrust force. Kuram et al. [68] formulated crude and refined sunflower based cutting fluids and used these vegetable based cutting fluids during drilling so that evaluated the performance of them measuring the thrust force and surface roughness. Sunflower based cutting fluid prepared using two different surfactants showed smaller force and surface roughness values as compared to using only one surfactant. Refined sunflower based cutting fluid gave lower surface roughness than crude sunflower based cutting fluid, while crude sunflower based cutting fluid showed lower thrust force than refined sunflower based cutting fluid. Belluco and De Chiffre [70] investigated the performance of vegetable based cutting fluids with determining the cutting force and power in the drilling, core drilling, reaming and tapping of AISI 316L stainless steel. Vegetable based cutting fluids could achieve equal or better efficiency than the reference commercial mineral oil in all operations [70]. Kuram et al. [71] investigated the effect of cutting fluids developed from raw and refined sunflower oil and two other commercial (vegetable and mineral based) cutting fluids on thrust force and surface roughness during drilling of AISI 304 stainless steel. Refined sunflower cutting fluid showed better or comparable performance to commercial vegetable based cutting fluid depending on cutting conditions. Vegetable based cutting fluids developed from refined sunflower oil performed better than that of semi-synthetic and mineral cutting fluids during drilling of AISI 304 austenitic stainless steel [72].

5.3.2 Turning

Khan and Dhar [12] investigated the effect of MQL by vegetable oil on cutting temperature, tool wear, surface roughness and dimensional deviation in turning of AISI 1060 steel. MQL reduced the cutting temperature, tool wear and surface roughness as compared to dry machining. The reduction of about 5–12 % at average cutting temperature using MQL by vegetable oil as compared to using dry machining was observed depending upon the levels of the process parameters (cutting speed, feed rate). Cutting forces decreased by about 5–15 % using MQL by vegetable oil. Ozcelik et al. [31] reported the experimental studies of sunflower and canola oils based cutting fluids including different percentage (8 and 12 %) of extreme pressure additive and two commercial cutting fluids (semi-synthetic and mineral based) in turning of AISI 304L stainless steel with respect to surface roughness, cutting force, feed force and tool wear. Experiments were also conducted at dry cutting conditions which caused rapid tool wear and fracture. Tool life below 200, 1,000 and 2,000 s were recorded under dry cutting, semi-synthetic and mineral based cutting fluids, respectively. The higher tool life in vegetable based cutting fluids was due to the fatty acid content. Canola based cutting fluids showed better performance than sunflower based cutting fluids because of different length of carbon chains. Canola oil has three carbons more in formulae and longer carbon chain outstand high cutting temperature, thus improving the surface protection. Moreover, the high viscosity of canola oil had a tendency to resist flow. This high viscosity provides more effective lubricating at the tool-chip interface, thus reduces the friction between the tool and workpiece and removes heat developed at the interface easily. High percentage of extreme pressure additive in vegetable based cutting fluids showed the higher surface roughness values. 8 % of extreme pressure additive included canola based cutting fluid performed better than the rest [31]. Higher rate of EP in sunflower and canola based cutting fluids reduced cutting and feed forces during turning of AISI 304L austenitic stainless steel, however the increment of EP rate affected surface roughness values negatively. As a result, mineral and semi-synthetic cutting fluids can be replaced by vegetable based cutting fluids in turning [16]. Ojolo et al. [73] used vegetable based cutting fluids (groundnut oil, coconut oil, palm kernel oil and shear butter oil) during turning of mild steel, aluminum and copper and measured cutting force. Although, it was found that the effects of vegetable based cutting fluids were material dependent, groundnut oil showed the best performance among the four vegetable based cutting fluids investigated. Xavior and Adithan [74] used coconut oil during turning of AISI 304 stainless steel and measured tool wear and surface roughness. The performance of coconut oil was compared with an emulsion and a neat cutting oil. They found that coconut oil reduced the tool wear and improved the surface finish. In another study they measured temperature and cutting force [75]. Coconut oil outperformed the other two cutting fluids (soluble oil and straight cutting oil) in terms of reducing the cutting force and temperature. Paul and Pal [76] investigated the performance of different types of cutting fluids (karanja oil, neem oil, conventional fluid) as compared to dry cutting condition during turning of mild steel. The use of vegetable based cutting fluid improved surface quality as compared to dry turning and conventional cutting fluid. They explained the lower temperature of neem vegetable oil than that of karanja vegetable oil with the lower viscosity of neem oil with respect to karanja oil.

5.3.3 Milling

Sales et al. [56] evaluated tool wear, surface roughness and burr formation when milling AISI 4140 steel with vegetable based cutting fluid applied by MQL technique using different flow rate (dry cutting, 50, 100, 150 and 200 ml/h). They stated that the vegetable cutting fluid efficiently accessed and remained at the chip-tool interface longer. This fact was due to the capability of vegetable oil' creating a thin film of molecular layer. This thin film improved boundary lubrication, thus decreased friction at the cutting interface. Increment in coolant flow rate tended to reduce tool wear, surface roughness and burr length. In milling of 15-5 precipitation-hardened martensitic stainless steel, Junior et al. [77] used four different cooling and lubrication conditions: flood of vegetable oil-based emulsion, low flow of neat vegetable oil, application of neat vegetable oil in a flow of compressed air (MQL) and dry cutting. The longest tool life was obtained with low flow of neat vegetable oil, followed by neat vegetable oil under MQL. Flood of vegetable oil-based emulsion showed worst performance. Kuram et al. [78] developed vegetable based cutting fluids from refined canola and sunflower oil which were used in milling of AISI 304 stainless steel and their tool wear and cutting force performances were compared to a commercial type semi-synthetic cutting fluid. Canola and sunflower based cutting fluids gave better performance than semi-synthetic cutting fluid.

5.3.4 Grinding

Oliveira and Alves [79] formulated new water based cutting fluid (sulfonate vegetable oil in water) able to meet both the grinding performance and environmental requirements. As a vegetable oil they selected sulfonate castor oil obtained from a plant called *mamona* in South America. The authors selected this oil owing to its abundance in South America and its stability. The new cutting fluid did not contain any banned products in its formula such as chlorine substances and nitrosamines. From the biodegradability test results it was concluded that the new cutting fluid was easily biodegradable, was not detrimental to the environment and its disposal could be easily made. The grinding experiments were conducted at SAE 8640 workpiece material. Cutting oil and semi-synthetic cutting fluid were selected as reference fluids so that the performance of the new cutting fluid is compared. In the grinding experiments, several dilutions (35, 45 and 70 % in volume) of this new cutting fluid were tested. The lowest and highest grinding ratios (G ratio-material removed volume/wheel worn volume) were measured using semi-synthetic cutting fluid and neat oil, respectively. The new cutting fluid concentrated at 45 % gave the closest performance to oil, i.e. high G ratio. The lowest surface roughness values were achieved with the new cutting fluid at the concentration of 45 % and the lowest force values were obtained when using new fluid at 45 % and the neat oil. As a result, new cutting fluid gave comparable performance to the obtained with neat oil. In an another study, they used new castor cutting fluid at 15, 21 and 32 % concentrations [80]. The lowest wheel wear (high G ratio) and the highest wheel wear were observed with cutting oil and semi-synthetic cutting fluid, respectively. The concentration of 21 % gave similar performance to the cutting oil. The lowest surface roughness was obtained at 21 % concentration. Alves and de Oliveira [17] used castor oil in grinding and found that new vegetable based cutting fluid was readily biodegradable. Wheel wear, grinding forces and surface roughness reduced when the new cutting fluid diluted at 45 % was used.

References

1. Willing A (2001) Lubricants based on renewable resources—an environmentally compatible alternative to mineral oil products. Chemosphere 43:89–98
2. Rao DN, Srikant RR (2006) Influence of emulsifier content on cutting fluid properties. Proc Inst Mech Eng, Part B: Eng Manuf 220:1803–1806
3. Srikant RR, Rao DN, Rao PN (2009) Influence of emulsifier content in cutting fluids on cutting forces, cutting temperatures, tool wear, and surface roughness. Proc Inst Mech Eng, Part J: Eng Tribol 223:203–209
4. Srikant RR, Rao DN, Rao PN (2009) Experimental investigations on the influence of emulsifier content in cutting fluids on machined surface. Proc Inst Mech Eng, Part J: Eng Tribol 223:195–201
5. Hong SY, Broomer M (2000) Economical and ecological cryogenic machining of AISI 304 austenitic stainless steel. Clean Prod Proc 2:157–166
6. Shokrani A, Dhokia V, Newman ST (2012) Environmentally conscious machining of difficult-to-machine materials with regard to cutting fluids. Int J Mach Tools Manuf. DOI: 10.1016/j.ijmachtools.2012.02.002
7. Abou-El-Hossein KA (2008) Cutting fluid efficiency in end milling of AISI 304 stainless steel. Ind Lubr Tribol 60(3):115–120
8. Fox NJ, Stachowiak GW (2007) Vegetable oil-based lubricants—a review of oxidation. Tribol Int 40:1035–1046
9. Abdalla HS, Patel S (2006) The performance and oxidation stability of sustainable metalworking fluid derived from vegetable extracts. Proc Inst Mech Eng, Part B: Eng Manuf 220:2027–2040
10. Krahenbuhl U (2002) Vegetable oil-based coolants improve cutting performance (Cutting Fluids). Tooling & Production, Nelson Publishing
11. Woods S (2005) Going green. Cutting Tool Eng 57(2):48–51
12. Khan MMA, Dhar NR (2006) Performance evaluation of minimum quantity lubrication by vegetable oil in terms of cutting force, cutting zone temperature, tool wear, job dimension and surface finish in turning AISI-1060 steel. J Zhejiang Univ Sci A 7(11):1790–1799
13. Carlsson AS (2006) Production of wax esters in crambe, Outputs from the EPOBIO Project, CPL Press, Newbury, Berks, p 60
14. Pettersson A (2007) High-performance base fluids for environmentally adapted lubricants. Tribol Int 40:638–645
15. Norrby T (2003) Environmentally adapted lubricants—where are the opportunities? Ind Lubr Tribol 55(6):268–274
16. Cetin MH, Ozcelik B, Kuram E, Demirbas E (2011) Evaluation of vegetable based cutting fluids with extreme pressure and cutting parameters in turning of AISI 304L by Taguchi method. J Clean Prod 19:2049–2056
17. Alves SM, Oliveira JFG (2008) Vegetable based cutting fluid-an environmental alternative to grinding process. In: 15th CIRP international conference on life cycle engineering, Syndey, pp 664–668

18. Wakabayashi T, Inasaki I, Suda S, Yokoto H (2003) Tribologic characteristics and cutting performance of lubricant esters for semi-dry machining. CIRP Ann-Manuf Techn 52:61–64
19. Rao PN (2009) Manufacturing technology: metal cutting & machine tools, vol II, 2nd ed., Tata McGraw-Hill Publishing, New Delhi
20. Mendes OC, Ávila RF, Abrão AM, Reis P, Davim JP (2006) The performance of cutting fluids when machining aluminium alloys. Ind Lubr Tribol 58(5):260–268
21. Machado AR, Wallbank J, Pashby IR, Ezugwu EO (1998) Tool performance and chip control when machining ti6A14 V and inconel 901 using high pressure coolant supply. Mach Sci Technol 2(1):1–12
22. Kaminski J, Alvelid B (2000) Temperature reduction in the cutting zone in water-jet assisted turning. J Mater Process Technol 106:68–73
23. Ezugwu EO, Bonney J (2004) Effect of high-pressure coolant supply when machining nickel-base, inconel 718 alloy with coated carbide tools. J Mater Process Technol 153–154:1045–1050
24. Ezugwu EO, Bonney J (2005) Finish machining of nickel-base inconel 718 alloy with coated carbide tool under conventional and high-pressure coolant supplies. Tribol Trans 48(1):76–81
25. Stanford M, Lister PM, Kibble KA (2007) Investigation into the effect of cutting environment on tool life during the milling of a BS970-080A15 (En32b) low carbon steel. Wear 262:1496–1503
26. Diniz AE, Micaroni R (2007) Influence of the direction and flow rate of the cutting fluid on tool life in turning process of AISI 1045 steel. Int J Mach Tools Manuf 47:247–254
27. Sharman ARC, Hughes JI, Ridgway K (2008) Surface integrity and tool life when turning inconel 718 using ultra-high pressure and flood coolant systems. Proc Inst Mech Eng, Part B: Eng Manuf 222:653–664
28. Nandy AK, Gowrishankar MC, Paul S (2009) Some studies on high-pressure cooling in turning of Ti-6Al-4 V. Int J Mach Tools Manuf 49:182–198
29. Davim JP, Sreejith PS, Gomes R, Peixoto C (2006) Experimental studies on drilling of aluminium (AA1050) under dry, minimum quantity of lubricant, and flood-lubricated conditions. Proc Inst Mech Eng, Part B: Eng Manuf 220:1605–1611
30. Sreejith P, Ngoi B (2000) Dry machining: machining of the future. J Mater Process Technol 101:287–291
31. Ozcelik B, Kuram E, Cetin MH, Demirbas E (2011) Experimental investigations of vegetable based cutting fluids with extreme pressure during of AISI 304L. Tribol Int 44:1864–1871
32. Ozcelik B, Kuram E, Simsek BT (2011) Comparison of dry and wet end milling of AISI 316 stainless steel. Mater Manuf Processes 26(8):1041–1049
33. Diniz AE, Júnior AB, Filho FT (2008) Tool wear damage caused by abundant emulsion in milling operation of PH stainless steel. J Braz Soc Mech Sci Eng 30(2):133–138
34. Diniz AE, Micaroni R (2002) Cutting conditions for finish turning process aiming: the use of dry cutting. Int J Mach Tools Manuf 42:899–904
35. Attanasio A, Gelfi M, Giardini C, Remino C (2006) Minimal quantity lubrication in turning: effect on tool wear. Wear 260:333–338
36. Su Y, He N, Li L, Iqbal A, Xiao M, Xu S, Qiu B (2007) Refrigerated cooling air cutting of difficult-to-cut materials. Int J Mach Tools Manuf 47:927–933
37. Vikram Kumar CHR, Ramamoorthy B (2007) Performance of coated tools during hard turning under minimum fluid application. J Mater Process Technol 185:210–216
38. Kelly JF, Cotterell MG (2002) Minimal lubrication machining of aluminium alloys. J Mater Process Technol 120:327–334
39. Braga DU, Diniz AE, Miranda GWA, Coppini NL (2002) Using a minimum quantity of lubricant (MQL) and a diamond coated tool in the drilling of aluminium-silicon alloys. J Mater Process Technol 122:127–138
40. Zeilmann RP, Weingaertner WL (2006) Analysis of temperature during drilling of ti6Al4 V with minimal quantity of lubrication. J Mater Process Technol 179:124–127
41. Heinemann R, Hinduja S, Barrow G, Petuelli G (2006) Effect of MQL on the tool life of small twist drills in deep-hole drilling. Int J Mach Tools Manuf 46:1–6

42. Tasdelen B, Thordenberg H, Olofsson D (2008) An experimental investigation on contact length during minimum quantity lubrication (MQL) machining. J Mater Process Technol 203:221–231
43. Costa ES, da Silva MB, Machado AR (2009) Burr produced on the drilling process as a function of tool wear and lubricant-coolant conditions. J Braz Soc Mech Sci Eng 31(1):57–63
44. Bhowmick S, Alpas AT (2008) Minimum quantity lubrication drilling of aluminium-silicon alloys in water using diamond-like carbon coated drills. Int J Mach Tools Manuf 48:1429–1443
45. Meena A, El Mansori M (2011) Study of dry and minimum quantity lubrication drilling of novel austempered ductile iron (ADI) for automotive applications. Wear 271:2412–2416
46. Varadarajan AS, Philip PK, Ramamoorthy B (2002) Investigation on hard turning with minimal cutting fluid application (HTMF) and its comparison with dry and wet turning. Int J Mach Tools Manuf 42:193–200
47. Diniz AE, Ferreira JR, Filho FT (2003) Influence of refrigeration/lubrication condition on SAE 52100 hardened steel turning at several cutting speeds. Int J Mach Tools Manuf 43:317–326
48. Dhar N, Islam M, Islam S, Mithu M (2006) The influence of minimum quantity of lubrication (MQL) on cutting temperature, chip and dimensional accuracy in turning AISI-1040 steel. J Mater Process Technol 171:93–99
49. Kamata Y, Obikawa T (2007) High speed MQL finish-turning of Inconel 718 with different coated tools. J Mater Process Technol 192–193:281–286
50. Sreejith PS (2008) Machining of 6061 aluminium alloy with MQL, dry and flooded lubricant conditions. Mater Lett 62:276–278
51. Jayal AD, Balaji AK (2009) Effects of cutting fluid application on tool wear in machining: interactions with tool-coatings and tool surface features. Wear 267(9–10):1723–1730
52. López de Lacalle LN, Angulo C, Lamikiz A, Sánchez JA (2006) Experimental and numerical investigation of the effect of spray cutting fluids in high speed milling. J Mater Process Technol 172:11–15
53. Wu C-H, Chien C-H (2007) Influence of lubrication type and process conditions on milling performance. Proc Inst Mech Eng, Part B: Eng Manuf 221:835–843
54. Iqbal A, Ning H, Khan I, Liang L, Dar NU (2008) Modeling the effects of cutting parameters in MQL-employed finish hard-milling process using D-optimal method. J Mater Process Technol 199:379–390
55. Thepsonthi T, Hamdi M, Mitsui K (2009) Investigation into minimal-cutting-fluid application in high-speed milling of hardened steel using carbide mills. Int J Mach Tools Manuf 49:156–162
56. Sales W, Becker M, Barcellos CS, Jr JL, Bonney J, Ezugwu EO (2009) Tribological behaviour when face milling AISI 4140 steel with minimum quantity fluid application. Ind Lubr Tribol 61(2):84–90
57. Fratila D, Caizar C (2011) Application of Taguchi method to selection of optimal lubrication and cutting conditions in face milling of $AlMg_3$. J Clean Prod 19(6–7):640–645
58. Tawakoli T, Hadad MJ, Sadeghi MH, Daneshi A, Stöckert S, Rasifard A (2009) An experimental investigation of the effects of workpiece and grinding parameters on minimum quantity lubrication–MQL grinding. Int J Mach Tools Manuf 49:924–932
59. Tawakoli T, Hadad MJ, Sadeghi MH, Daneshi A, Sadeghi B (2011) Minimum quantity lubrication in grinding: effects of abrasive and coolant–lubricant types. J Clean Prod 19:2088–2099
60. Hadad MJ, Tawakoli T, Sadeghi MH, Sadeghi B (2012) Temperature and energy partition in minimum quantity lubrication–MQL grinding process. Int J Mach Tools Manuf 54–55:10–17
61. Mao C, Tang X, Zou H, Zhou Z, Yin W (2012) Experimental investigation of surface quality for minimum quantity oil–water lubrication grinding. The Int J Adv Manuf Technol 59:93–100
62. Rahman M, Senthil Kumar A, Salam MU (2002) Experiment evaluation on the effect of minimal quantities of lubricant in milling. Int J Mach Tools Manuf 42:539–547

63. Aoyama T (2002) Development of a mixture supply system for machining with minimal quantity lubrication. Ann CIRP 51(1):289–292
64. Sharma VS, Dogra M, Suri NM (2009) Cooling techniques for improved productivity in turning. Int J Mach Tools Manuf 49:435–453
65. Varadarajan AS, Philip PK, Ramamoorthy B (2002) Investigations on hard turning with minimal cutting fluid application (HTMF) and its comparison with dry and wet turning. Int J Mach Tools Manuf 42:193–200
66. Obikawa T, Kamata Y, Shinozuka J (2006) High-speed grooving with applying MQL. Int J Mach Tools Manuf 46:1854–1861
67. Belluco W, De Chiffre L (2004) Performance evaluation of vegetable-based oils in drilling austenitic stainless steel. J Mater Process Technol 148:171–176
68. Kuram E, Ozcelik B, Demirbas E, Şık E, Tansel IN (2011) Evaluation of new vegetable-based cutting fluids on thrust force and surface roughness in drilling of AISI 304 using Taguchi method. Mater Manuf Processes 26(9):1136–1146
69. Rahim EA, Sasahara H (2011) A study of the effect of palm oil as MQL lubricant on high speed drilling of titanium alloys. Tribol Int 44:309–317
70. Belluco W, De Chiffre L (2001) Testing of vegetable-based cutting fluids by hole making operations. Lubr Eng 57(1):12–16
71. Kuram E, Ozcelik B, Demirbas E, Şık E (2010) Effects of the cutting fluid types and cutting parameters on surface roughness and thrust force. Proceedings of the world congress on engineering, London, pp 1312–1315
72. Ozcelik B, Kuram E, Demirbas E, Şık E (2011) Optimization of surface roughness in drilling using vegetable-based cutting oils developed from sunflower oil. Ind Lubr Tribol 63(4):271–276
73. Ojolo SJ, Amuda MOH, Ogunmola OY, Ononiwu CU (2008) Experimental determination of the effect of some straight biological oils on cutting force during cylindrical turning. Revista Matéria 13(4):650–663
74. Xavior MA, Adithan M (2009) Determining the influence of cutting fluids on tool wear and surface roughness during turning of AISI 304 austenitic stainless steel. J Mater Process Technol 209(2):900–909
75. Xavior MA, Adithan M (2010) Evaluating the performance of cutting fluids in machining of AISI 304 austenitic stainless steel. Int J Mach Mach Mater 7(3/4):244–259
76. Paul S, Pal PK (2011) Study of surface quality during high speed machining using eco-friendly cutting fluid. Mach Technol Mater 11:24–28
77. Junior AB, Diniz AE, Filho FT (2009) Tool wear and tool life in end milling of 15–5 PH stainless steel under different cooling and lubrication conditions. Int J Adv Manuf Technol 43:756–764
78. Kuram E, Simsek BT, Ozcelik B, Demirbas E, Askin S (2010) Optimization of the cutting fluids and parameters using Taguchi and ANOVA in milling. Proceedings of the world congress on engineering, London, pp 1292–1296
79. Oliveira JFG, Alves SM (2006) Development of environmentally friendly fluid for CBN grinding. CIRP Ann-Manuf Technol 55:343–346
80. Alves SM, de Oliveira JFG (2006) Development of new cutting fluid for grinding process adjusting mechanical performance and environmental impact. J Mater Process Technol 179:185–189

Environmental Friendly Joining of Tubes

Luis M. Alves and Paulo A. F. Martins

Abstract This chapter introduces an innovative and environmental friendly joining technology for connecting tubes and fixing tubes to sheet panels that can successfully eliminate existing technologies based on fasteners, welding and structural adhesive bonding in standard or user made types of connections. The proposed joining technology is built upon compression beading of thin-walled tubes at room temperature and in some cases may also involve flaring the tube ends. In terms of deformation mechanics, the aim of this chapter is to identify the main parameters, to diagnose possible sources of failure and to understand the route for selecting the most appropriate operative conditions for obtaining sound joints between tubes and between tubes and sheet panels. The joining technology to be presented in this chapter is suitable for mass production and copes with the growing agile manufacturing trends and sustainability issues requiring flexibility, short life-cycles, very short development and production lead times. Emphasis is placed on tooling systems, finite element modeling and experimentation with the objective of presenting a wide range of engineering applications involving custom sizes and dissimilar materials (e.g. metals and polymers) and showing the flexibility and cost effectiveness of the proposed technology.

1 Introduction

Conventional technologies for joining tubes to sheet panels are widely utilized in transportation vehicles, lightweight structures and process piping, among other applications.

There are basically three different ways to joint tubes and sheets together; mechanical fixing with fasteners (nuts and bolts or rivets), welding and structural

L. M. Alves · P. A. F. Martins (✉)
IDMEC, Instituto Superior Técnico, Technical University of Lisbon,
Av. Rovisco Pais, 1049-001 Lisbon, Portugal
e-mail: pmartins@ist.utl.pt

J. P. Davim (ed.), *Green Manufacturing Processes and Systems*, Materials Forming, Machining and Tribology, DOI: 10.1007/978-3-642-33792-5_3,

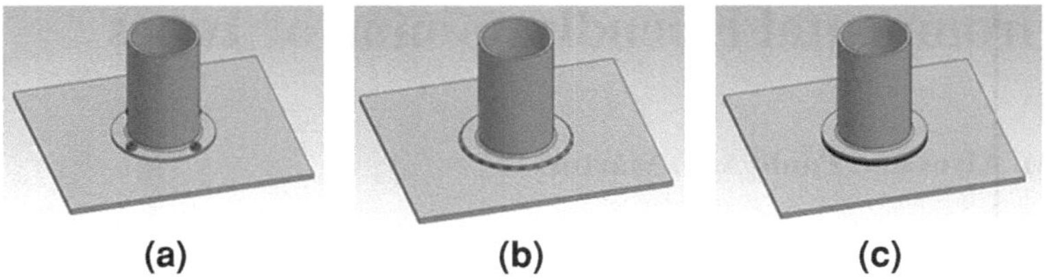

Fig. 1 Conventional technologies for joining tubes to sheet panels by means of **a** mechanical fixing with fasteners, **b** welding and **c** structural adhesive bonding

adhesive bonding (Fig. 1). These technologies are very different of one another and their universe of applicability is limited by aesthetic, physical, chemical and mechanical requirements.

The utilization of mechanical fasteners usually suffers from unwanted aesthetic features, corrosion problems or functional difficulties related to the maximum effort that nuts, bolts and rivets can support safely.

Welding has similar problems to those mentioned for mechanical fasteners plus specific difficulties related to the heat-cooling cycles, to the weldability of materials namely, dissimilar materials (e.g. joining steel or aluminium tubes to aluminium or copper sheets), and to the production of undesirable fumes and smokes in fabrication.

Structural adhesive bonding offers engineers the possibility of joining different types of materials while improving aesthetics by avoiding rivets and bolt heads. However, adhesives require careful preparation of the surfaces where they are to be applied and may experience significant decrease in performance over time and in the presence of hostile environmental conditions (e.g. prolonged expose in moist environments).

In addition to what was mentioned before it is worth notice that conventional technologies currently utilized for joining tubes to sheet panels are not entirely adequate for large batch sizes due to high set-up and production run times.

Conventional technologies for connecting tubes (also known as 'tube branching technologies') are widely utilized in plumbing, air conditioned, refrigeration, process piping and lightweight structures, among other applications. Figure 2 shows five different ways to connect tubes to each other.

The most well-known technologies are based on commercially available tee fittings, saddle adapters and weld-o-lets for standard geometries and materials, such as carbon steel, stainless steel and copper (Figs. 2a–c). A standard tee fitting (Fig. 2a) has three welds; two in the main tube and one in the branch tube. Saddle adapters or weld-o-lets (Fig. 2b and c) also need to be brazed or welded to the main tube over a pre-cut hole and the attachment to the branch tube is made through a weld or a threated connection.

The advantage of connecting tubes by means of commercially available solution is that it is a cost effective option whenever the quantity of joints to be performed is significant. The problem with commercially available solutions is that

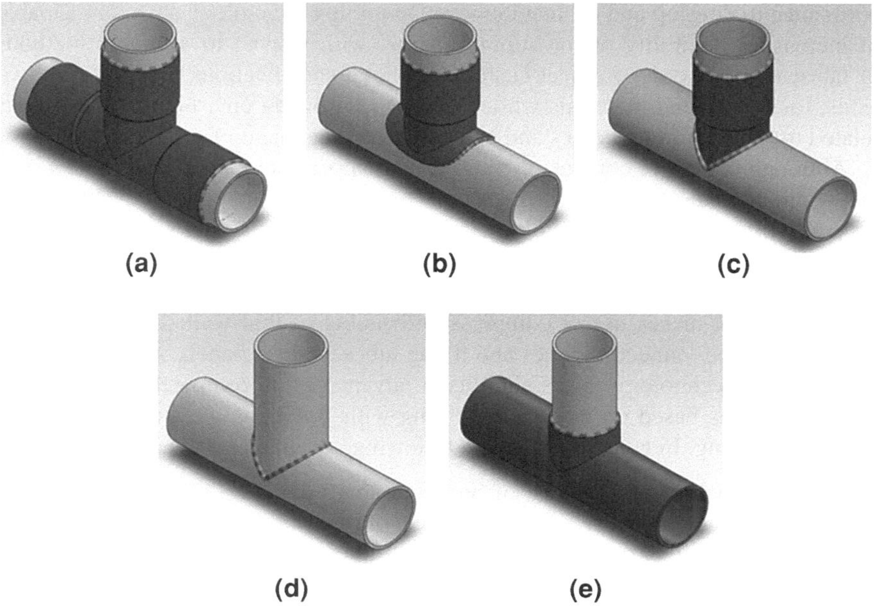

Fig. 2 Conventional tube branching by means of **a** tee fittings, **b** saddle adapters, **c** weld-o-lets, **d** nozzle-welds and **e** spin-forming

connections are limited by industry standards and always require cutting and preparation of tube ends, welding or brazing and, in some cases, quality inspection of the welds. Therefore, because commercially available solutions are not appropriate for obtaining connections with non-standard geometries, there are alternative tube branching methods that take advantage of the user's ability to fabricate its own connections (Fig. 2d and e).

The nozzle-weld is the most commonly fabricated connection but other solutions based, for instance, on spin-forming are also frequently employed. Nozzle-weld connections (Fig. 2d) require cutting a hole in the main tube, shaping a contoured end in the branch tube to match the diameter of the main tube and welding along the contour. Spin-forming (Fig. 2e) also requires cutting a hole in the main tube. The difference is that material around that hole is subsequently shaped into a tee fitting where the branch tube will be brazed or welded.

From what was mentioned before, it can be concluded that despite differences in conventional tube branching technologies they all require the utilization of similar materials with good weldability, all give rise to heat-cooling cycles due to welding or brazing, all require filling materials and all are responsible for producing undesirable fumes and smokes in fabrication. This not only causes difficulties in dimensional accuracy and quality but also gives rise to environmental problems in production lines.

The facts and reasons behind the utilization of conventional joining technologies for connecting tubes and fixing tubes to sheets allow us to conclude on the pertinence to develop and evaluate innovative joining technologies that are capable of increasing flexibility of the supply chains (with respect to available methods of fabrication) in order to meet customer-defined specifications regarding dimensions, tolerances and materials while addressing growing environmental concerns related to fabrication processes and end-of-life management of products.

Although the above conclusion may, at first, appear obvious it represents a serious technological challenge for the current state-of-the art of joining due to the necessity of developing a technology that is capable of satisfying so many requirements along with high quality and low operative costs.

This chapter presents a cost competitive and environmental friendly joining technology that makes use of compression beading of thin-walled tubes, at room temperature, for connecting tubes and fixing tubes to sheet panels. As will be seen, the proposed technology offers significant advantages as compared with conventional processes based on mechanical fixing with fasteners, welding or structural adhesive bonding. In general terms, the following advantages can be invoked:

(i) Flexible technology capable of handling small, medium or large batch sizes with different geometries and high levels of repeatability in production line;
(ii) Environmentally friendly technology that allows savings in raw material and eliminates filler materials and shielding gases;
(iii) Energy saving technology that eliminates heat-cooling cycles as well as heat affected zones and residual stresses in the regions of the tubes and sheet panels that are joined together;
(iv) Value added technology that is capable of connecting tubes and fixing tubes to sheet panels made of dissimilar materials;
(v) Cost-efficient technology that requires low amount of capital investment because it can be design to operate with existing machine-tools.

Pioneering research work in plastic instability of tubes due to axial compression loading was performed by Alexander [1] and Allan [2] who concluded that collapse mechanisms by local buckling are the result of inward and outward movements of the tube wall. This view of plastic deformation was later enhanced by Rosa et al. [3] who modelled the contact between successive instability waves as well as the progressive changes taking place at the contact region between the tube and the parallel flat platens.

On contrary to plastic instability of tubes by local buckling, little research has been conducted on compression beading of thin walled tubes using a die. Besides the work by Gouveia et al. [4], who review the deformation mechanisms and establish feasibility intervals as a function of the major operative parameters, one of the very few publications in the field is due to Al-Qureshi et al. [5] who studied compression beads formed by elastomers. As a result of this, current design rules for compression beading of thin-walled tubes using a die are mainly derived from accumulated experience and technical data available in specialized literature [6].

In case of connecting tubes and fixing tubes to sheet panels by means of plastic deformation, at room temperature, the only systematic research works available in the literature are those recently published by Alves et al. [7, 8] that will be appropriately refereed throughout the presentation. Aims and objectives of this chapter are threefold: (i) to characterize the occurrence of local buckling in thin-walled tubes subjected to axial loading, in order to provide the necessary background to understand the physics of compression beading, (ii) to propose a new environmental joining technology for connecting tubes and fixing tubes to sheet panels based on compression beading and (iii) to demonstrate the feasibility of the proposed technology by showing examples of joined parts.

The chapter also reviews the methodology utilized in mechanical testing of materials and the fundamentals of the finite element flow formulation that assisted authors in the conception and design of the tooling systems that are commonly utilized in the proposed joining technology based on tube forming.

2 Mechanical Joining by Tube Forming

2.1 Plastic Instability

Plastic instability of tubes under axial compression is characterized by two fundamental modes of deformation; global buckling and local buckling (Fig. 3). Global buckling takes place when a tube fails as a whole (that is, like a column) and is expected to occur when the tube is long and has relatively thick walls. Local buckling leads to the development of axisymmetric wrinkles (or compression beads) along the tube and usually takes place when a tube, either short or long, has thin walls.

The mechanical characterization of global buckling follows the elementary theory of columns due to Euler, and later improved by Engesser and Shanley [9] on the basis of the tangent-modulus formula,

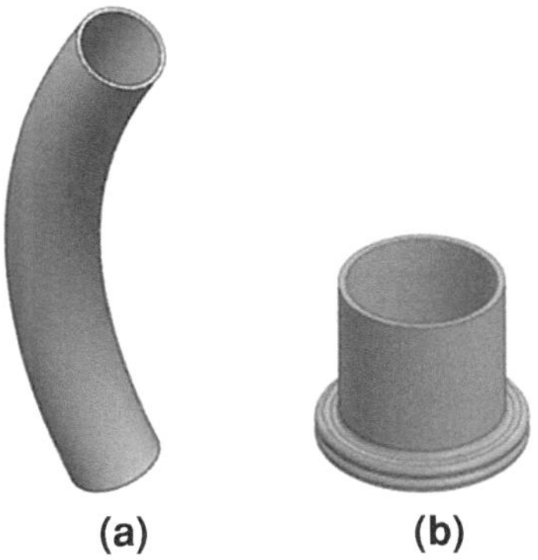

Fig. 3 Plastic instability of tubes under axial compression. **a** Global buckling and **b** local buckling with formation of axisymmetric compression beads

$$\sigma_{cr} = 0.5C\pi^2 E_t \left(\frac{r_0}{l}\right)^2 \tag{1}$$

where l is the length, r_0 is the reference radius, C is a parameter related with the end conditions ($C = 4$, when both ends are fixed) and E_t is the tangent modulus of the material defined as the slope $d\sigma / d\varepsilon$, of the stress–strain curve in the plastic range at the current value of stress $\sigma = F / A$.

The critical axial compressive stress σ_{cr} that originates local buckling (or wrinkling) on a thin-walled tube is calculated from the work of Timoshenko [10],

$$\sigma_{cr} = \frac{1}{\sqrt{3 \cdot (1 - \nu^2)}} E_t \frac{t}{r_0} \tag{2}$$

where, t is the thickness and ν is the Poisson ratio. The above Eq. 2 assumes concentric and uniform loading conditions and, therefore, can only be applied to the development of axisymmetric local buckling in thin-walled tubes under axial compression loading.

2.2 Compression Beading

Compression beading is a tube forming operation that makes use of local buckling plastic instability mechanisms. The operation is accomplished by forcing one tube end towards the other (or the two tube ends towards one another) while leaving a gap opening in-between the dies that support/hold the tube ends (l_{gap} refers to the initial gap height). As the upper die compresses the tube it collapses at the gap opening creating the required bead Fig. 4a, b.

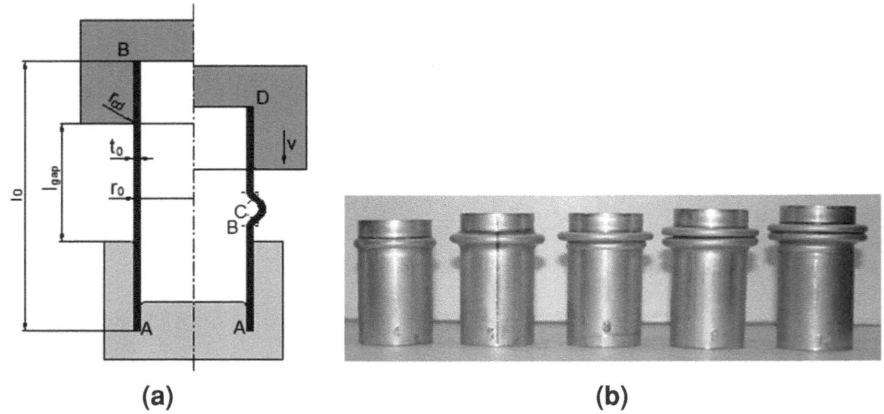

Fig. 4 **a** Schematic representation of the compression beading of thin-walled tubes and **b** photographs of compression beads produced in aluminium AA6060-O tubes with different operative parameters

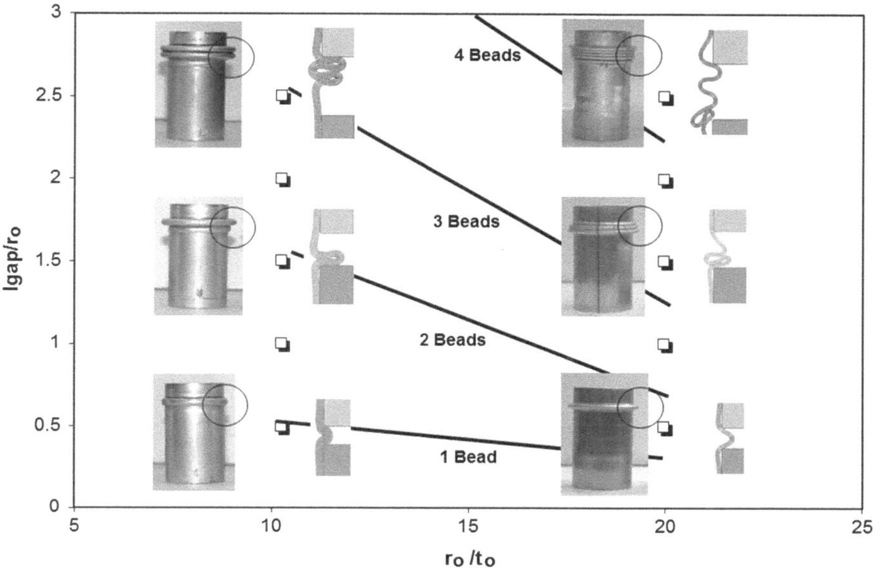

Fig. 5 Formability diagram of the compression beading of thin-walled aluminium AA6060-O tubes. *Note* The enclosed details of the cross section were obtained by finite element analysis

The main operative parameters of compression beading can be identified as the ratio between the reference radius and the thickness of the tube wall r_0/t_0 and the ratio between the initial gap height and the reference radius of the tube l_{gap}/r_0 (also known as the 'slenderness ratio').

Figure 5 shows the combined influence of both ratios in the compression beading of aluminium AA6060-O tubes. As seen, by varying the first parameter r_0/t_0, it is possible to divide material flow into two different groups. For high values of r_0/t_0 tube beads are steadily formed both outward and inward whereas, for decreasing values of r_0/t_0, material flow gradually changes to became outward dominant and, as a consequence, tube beads start to be formed mainly outward.

The workability diagram included in Fig. 5 presents several regions that were established by defining the onset number of tube beads for each combination of process parameters, l_{gap}/r_0 and r_0/t_0. The diagram helps understanding and predicting the number of tube beads in a compression beading operation and provides key in hand information for designing tooling systems for connecting tubes and fixing tubes to sheet panels.

In general terms, it may be concluded that the number of tube beads increases with the initial gap height and decreases with the wall thickness. However, caution and further analysis are necessary whenever the design specifications require tube beads to form predominantly outward. In fact, only using low values of r_0/t_0 it is likely to achieve outwardly dominant material flow conditions. Finally, it is worth

Fig. 6 Typical modes of deformation that may develop during asymmetric compression beading of thin-walled commercial S460MC tubes with different slenderness ratios

notice that besides the number of tube beads also the width of these beads can be slightly altered by setting the initial gap height. In fact, because the gap height decreases during deformation, each time a new bead starts to form it is necessary to provide it with a remaining compression distance big enough for its proper development (refer to pictures included in Fig. 5 showing several 'complete' and 'incomplete' tube beads).

Besides axisymmetric compression beads, which are naturally formed by local buckling during successive in-plane instability waves, there is also the possibility of artificially triggering the development of out-of-plane instability waves between contoured dies. Out-of-plane instability waves lead to the development of asymmetric compression beads and are necessary for connecting tubes to each other by means of the joining technology that will be proposed in the next section.

Figure 6 shows the influence of the slenderness ratio of the tube l_{gap}/r_0 on the development of asymmetric compression beads. The leftmost sample in the figure was unable to develop a compression bead because the initial gap opening had a value below the minimum threshold that is necessary to ensure local buckling by plastic instability mechanisms.

2.3 Tooling Systems

The proposed environmental friendly technology for joining tubes to sheet panels is schematically described in Fig. 7 for two different operative setups; two-stage axisymmetric compression beading (Fig. 7a) and axisymmetric compression beading followed by tube end flaring (Fig. 7b).

In case of Fig. 7a, clamping the tube to the sheet panel is ensured by means of two consecutive operations that promote collapse of the tube under local buckling

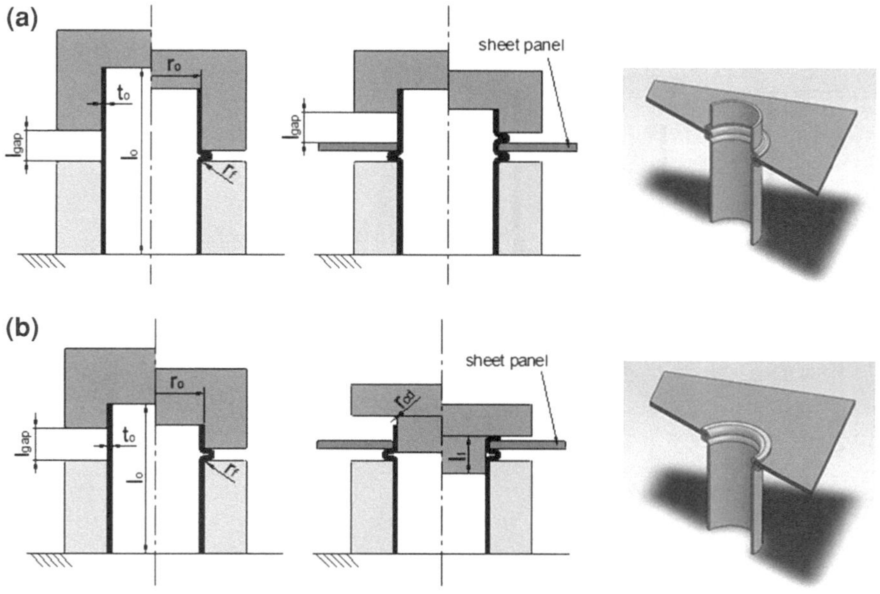

Fig. 7 Joining tubes to sheet panels. **a** Two-stage axisymmetric compression beading and **b** axisymmetric compression beading followed by tube end flaring

and give rise to axisymmetric compression beads. This mode of deformation is ensured by forcing the upper against the bottom tube end while leaving a gap opening in-between the dies that support/hold the tube.

In case of Fig. 7b, compression beading promotes collapse of the tube under local buckling and gives rise to an axisymmetric bead. Subsequent flaring clamps the tube end to the sheet panel placed on top of the bead. This mode of deformation is accomplished by compressing the upper free end of the tube with an appropriate radiused punch in order to expand material outwards and form a single-lap flange.

Figure 8a illustrates both before and after stages of the tooling system utilized for connecting tubes. As seen, asymmetric compression beading is accomplished by forcing one tube end towards the other while leaving a gap opening in-between the dies that support and hold the tubes. As the upper die compresses the tube it collapses at the gap opening creating the required asymmetric bead. The utilization of two opposite asymmetric compression beads allows connecting (locking) the tubes by plastic deformation at room temperature and is the basis for the new proposed tube branching technology.

In case the connection between tubes is to be performed at the tube end, the proposed joining technology needs to be modified in order to combine asymmetric compression beading with tube end flaring (Fig. 8b).

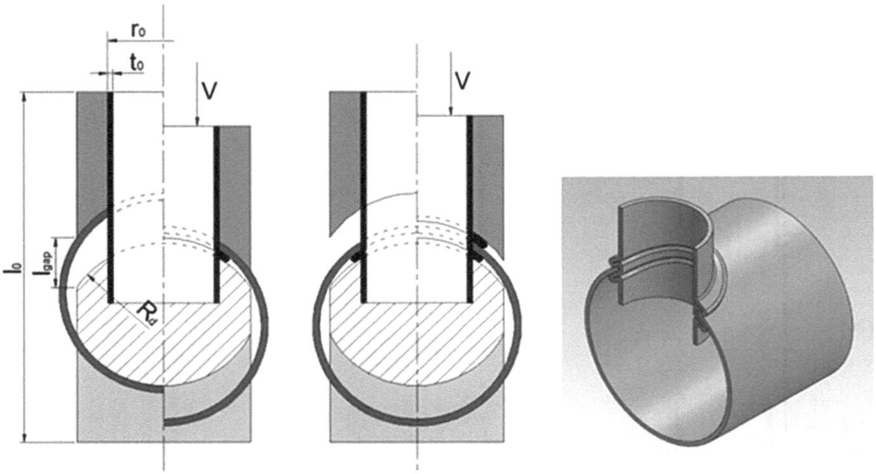

(a)

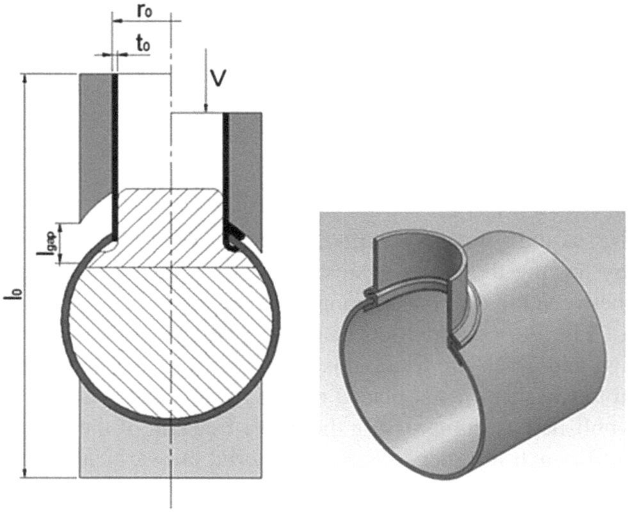

(b)

Fig. 8 Connecting tubes. **a** Two-stage asymmetric compression beading and **b** asymmetric compression beading followed by tube end flaring

3 Mechanical Characterization

3.1 Stress–Strain Curve

The stress–strain behaviour of the tubes can be determined by means of tensile and stack compression tests carried out at room temperature [11].

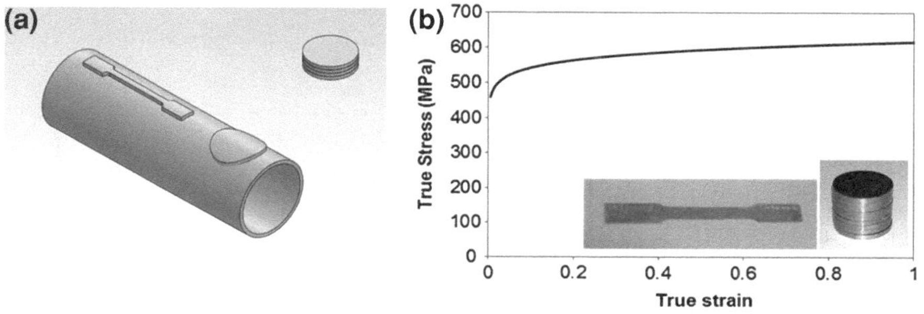

Fig. 9 **a** Schematic picture showing a tensile test specimen and a disc from a stack compression test specimen being removed from a tube. **b** Stress–strain curve of commercial S460MC tubes

The specimens utilized in the tensile tests were machined from the supplied tube stock while the specimens utilized in the stack compression tests were assembled by pilling up circular discs cut from the tube stock by a hole saw (Fig. 9).

The tests must be performed at room temperature on mechanical testing machines with cross-head speeds equal to those utilized in the joining processes. The effects of temperature, strain rate and anisotropy on material flow behaviour can generally be neglected.

3.2 Critical Instability Force

The critical instability force F_{cr} for the occurrence of axisymmetric local buckling in tubes subjected to axial loading is experimentally determined by compressing tubular specimens between flat parallel dies.

Figure 10 shows the force recorded during the experiments as a function of the displacement of the upper flat compression platen for commercial S460MC (carbon steel) tubes with a reference radius $r_0 = 16$ mm, wall thickness $t_0 = 1.5$ mm and initial length $l_0 = 100$ mm. As seen in the figure, the load increases sharply from zero and local buckling occurs upon reaching a critical experimental value $F_{cr} \cong 93.5$ kN. Subsequent deformation of the tubes results in additional axisymmetric compression beads during successive in-plane instability waves and to the development of a cyclic peak-to-peak load–displacement evolution.

The experimental value of the critical instability force compares well with the theoretical predictions based on Eq. 2 ($F_{cr} = 95$kN).

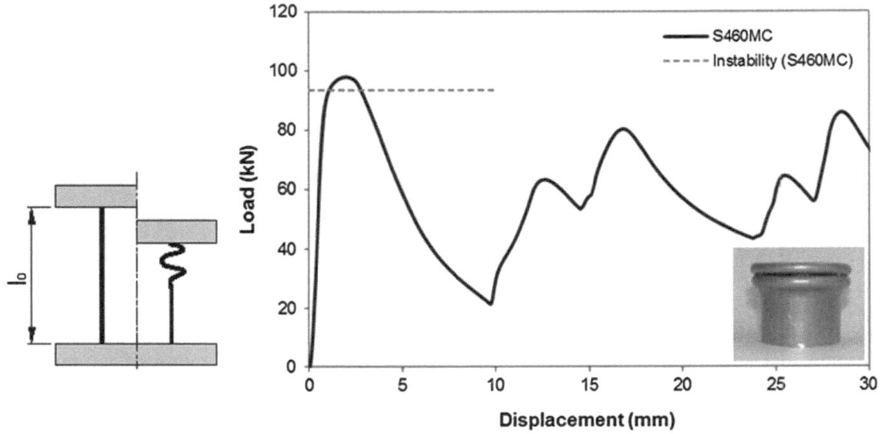

Fig. 10 Schematic picture showing a tube being compressed between flat parallel dies and load–displacement curve and critical instability force (refer to *dashed horizontal line*) for the occurrence of axisymmetric local buckling during axial compression of thin-walled S460MC tubes

4 Finite Element Modelling Background

4.1 Finite Element Flow Formulation

The finite element flow formulation giving support to the computer program i-form that assisted authors in the conception and design of the tooling systems of the proposed joining technology is built upon the following weak variational form (expressed in terms of the arbitrary variation in the velocity) [12],

$$\delta\,\Pi = \int_V \bar{\sigma}\,\delta\dot{\bar{\varepsilon}}\,dV + K\int_V \dot{\varepsilon}_V\,\delta\dot{\varepsilon}_V\,dV - \int_{S_T} t_i\,\delta u_i\,dS = 0 \tag{3}$$

where V is the control volume limited by the surfaces S_U and S_T where velocity and traction are prescribed respectively and K is a large positive constant penalising the volumetric strain rate component $\dot{\varepsilon}_v$ in order to enforce incompressibility.

The utilization of the finite element flow formulation based on the penalty function approach (also known as the 'irreducible flow formulation') offers the advantage of preserving the number of independent variables, because the average stress σ_m can be computed after the solution is reached through,

$$\sigma_m = K\,\dot{\varepsilon}_v \tag{4}$$

The effective stress and the effective strain rate are defined, respectively by,

$$\bar{\sigma} = \sqrt{\frac{3}{2}\sigma'_{ij}\sigma'_{ij}} \tag{5}$$

$$\dot{\bar{\varepsilon}} = \sqrt{\frac{2}{3}\dot{\varepsilon}'_{ij}\,\dot{\varepsilon}'_{ij}} \tag{6}$$

where σ'_{ij} is the deviatoric stress tensor and $\dot{\varepsilon}'_{ij}$ the deviatoric strain-rate tensor.

The spatial discretization of the weak variational form (Eq. 3) by means of M finite elements with constant pressure interpolation, linked through N nodal points results in the following set of nonlinear equations [12, 13],

$$\sum_{m=1}^{M}\left\{\int_{V^m}\frac{\bar{\sigma}}{\dot{\bar{\varepsilon}}}\delta\mathbf{v}^T\mathbf{K}\,\mathbf{v}\,dV^m + K^m\int_{V^m}\delta\mathbf{v}^T\mathbf{C}^T\mathbf{B}\mathbf{v}\mathbf{C}^T\mathbf{B}\,dV^m - \int_{S_T^m}\delta\mathbf{v}^T\mathbf{N}\mathbf{T}\,dS^m\right\} = 0 \tag{7}$$

which can be written in the following compact form,

$$\sum_{m=1}^{M}\left\{\left[\bar{\sigma}\ \mathbf{P} + K^m\mathbf{Q}\right]\ \{\mathbf{v}\} = \{\mathbf{F}\}\right\} \tag{8}$$

where,

$$\mathbf{P} = \int_{V^m}\frac{1}{\dot{\bar{\varepsilon}}_{n-1}}\mathbf{K}\,dV^m \tag{9}$$

$$\mathbf{K} = \mathbf{B}^T\mathbf{D}\,\mathbf{B} \tag{10}$$

$$\mathbf{Q} = \int_{V^m}\mathbf{C}^T\mathbf{B}\mathbf{C}^T\mathbf{B}\,dV^m \tag{11}$$

$$\mathbf{F} = \int_{S_T^m}\mathbf{N}\mathbf{T}\,dS^m \tag{12}$$

The symbol **N** denotes the matrix containing the shape functions of the finite element, **B** is the velocity-strain rate matrix, **C** is the matrix form of the Kronecker symbol and **D** is the matrix relating the deviatoric stresses with the strain rates according to the rate-form of the Levy–Mises constitutive equations.

The nonlinear set of Eq. 8 derived from the finite element flow formulation based on the penalty function approach, can be efficiently solved by a numerical technique resulting from the combination between the direct iteration and the Newton–Raphson methods.

The direct iteration method, which considers the Levy–Mises constitutive equations to be linear (and therefore constant) during each iteration, is to be preferentially utilized for generating the initial guess of the velocity field required by the

Newton–Raphson method. The Newton–Raphson method is an iterative procedure based on a Taylor linear expansion of the residual force vector $\mathbf{R}(\mathbf{v})$ of Eq. 8,

$$\mathbf{R}^n = \sum_{m=1}^{M} \{[\bar{\sigma}\,\mathbf{P} + K^m \mathbf{Q}]^n \,\{\mathbf{v}\}^n - \{\mathbf{F}\}^n\} \tag{13}$$

near the velocity estimate at the previous iteration,

$$\mathbf{R}\left(\mathbf{v}^n\right) \cong \mathbf{R}^n = \mathbf{R}^{n-1} + \left[\frac{\partial \mathbf{R}}{\partial \mathbf{v}}\right]_{n-1} \Delta\mathbf{v}^n = \mathbf{0} \tag{14}$$

where $\Delta\mathbf{v}$ is the first order correction of the velocity field, the symbol n denotes the current iteration number,

$$\{\mathbf{v}\}^n = \{\mathbf{v}\}^{n-1} + \alpha\,\{\Delta\mathbf{v}\}^n \qquad \alpha \in\,]0, 1] \tag{15}$$

and α is a parameter that controls the magnitude of the velocity correction term $\Delta\mathbf{v}$. This procedure is only conditionally convergent, but converges quadratically in the vicinity of the exact solution.

The aforementioned numerical methods are designed in order to minimize the residual force vector $\mathbf{R}(\mathbf{v})$ to within a specified tolerance. Control and assessment is performed by means of appropriate convergence criteria [13].

4.2 Contact and Friction

Contact with friction between workpiece and tooling requires tubes and sheets to be discretized by hexahedral (or quadrilateral) elements and dies (treated as rigid) to be discretized by spatial triangular (or linear) elements.

In case of three-dimensional modelling, the contact algorithm utilized by the finite element computer program i-form is based on node-to-triangle contact as illustrated in Fig. 11 by a workpiece node contacting a triangular element of the tool. Boundary nodes, like P in Fig. 11, are analysed for each triangular surface element of the tool. The orthogonal projection P_* of node P to the plane spanned by the triangle is calculated.

Figure 11b shows an example of the orthogonal projection being inside the considered triangular element, which is one of the conditions for being in contact. Figure 11c shows an example of the orthogonal projection lying outside, and hence node p and this triangular element is not in contact. The evaluation of whether or not the projection lies inside the triangle is based on a comparison of the total area of the triangle $A_{P_1 P_2 P_3}$ and the area sum $A_1 + A_2 + A_3$ of the triangles spanned by the projection point and two of the triangle vertices. If the point is inside, the two areas are identical. Another condition for being in contact is that

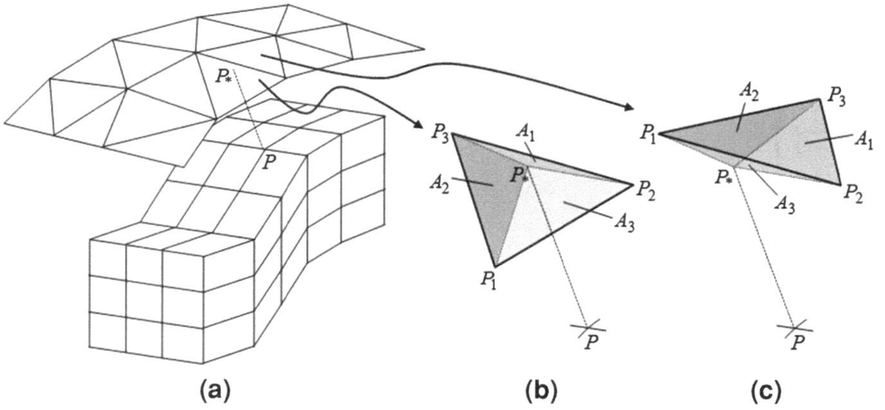

Fig. 11 Contact between the hexahedral mesh of the workpiece and the triangular surface mesh of a rigid tool. **a** Node P and its projection P_* on a triangular element, **b** normal projection P_* of node P lying inside triangular surface element and **c** normal projection P_* of node P lying outside triangular surface element

the distance between P and P_* is less than a specified value in order to avoid nodes far from the tool to be considered in contact.

The time increment Δt that is necessary for a nodal point P to get in contact with the tools is evaluated explicitly such that the increment of time needed for each of the potential nodes to get in contact with a tool is calculated according to,

$$\Delta t_p = \frac{\overline{PP_*}}{u_P - u_{P_*}} \tag{16}$$

where the denominator is the normal velocity difference between the candidate node P and its projection P_* on the tool, which if it is negative corresponds to an increasing gap and in that case it is discarded as a candidate. Among the candidates, the minimum time Δt_P from Eq. 16 is decisive for the following time increment. If the time step is larger than the minimum time for a contact point to arise, it is split to $\Delta t = \Delta t_P^{\min}$. All points getting in contact to the tools within a specified tolerance in the following step are projected to the tool and assigned boundary conditions to enforce that the projected points follow the movement of the tool.

Assuming friction at the contact interface between workpiece and tooling to be a traction boundary condition and the additional power consumption term to be modelled through the utilization of the law of constant friction $\tau_f = mk$, where k is the shear yield stress, the discretization of the weak variational in Eq. 7 can be modified as follows [13].

$$\sum_{m=1}^{M}\left\{\begin{array}{l}\int\limits_{V^m}\frac{\bar{\sigma}}{\dot{\bar{\varepsilon}}}\mathbf{K}\,\mathbf{v}\,dV^m + K^m\int\limits_{V^m}\mathbf{C}^T\mathbf{B}\,\mathbf{v}\,\mathbf{C}^T\mathbf{B}\,dV^m - \int\limits_{S_T^m}\mathbf{N}\,\mathbf{T}\,dS^m \\ \qquad + \int\limits_{S_{FR}^m} mk\frac{2}{\pi}\mathbf{N}\,\arctan\left[\frac{\mathbf{N}\mathbf{v}_r}{v_0}\right]dS^m\end{array}\right\} = 0 \tag{17}$$

where $\mathbf{v}_r$ denotes the relative sliding velocity (in matrix form) between the workpiece and tooling.

The approximation of the frictional stress $\tau_f = mk$ by an arctangent function of the relative sliding velocity u_r eliminates the sudden change of direction of the frictional stress at the neutral point,

$$\tau_f = mk\left\{\frac{2}{\pi}\arctan\left(\frac{|u_r|}{u_0}\right)\right\}\frac{u_r}{|u_r|} \tag{18}$$

where u_0 is an arbitrary value within the range 10^{-3} to 10^{-4} in order to avoid numerical difficulties.

4.3 Modelling Conditions

The numerical evaluation of the volume integrals included in Eq. 8 is performed by means of a standard discretization procedure. In case of rotational symmetry

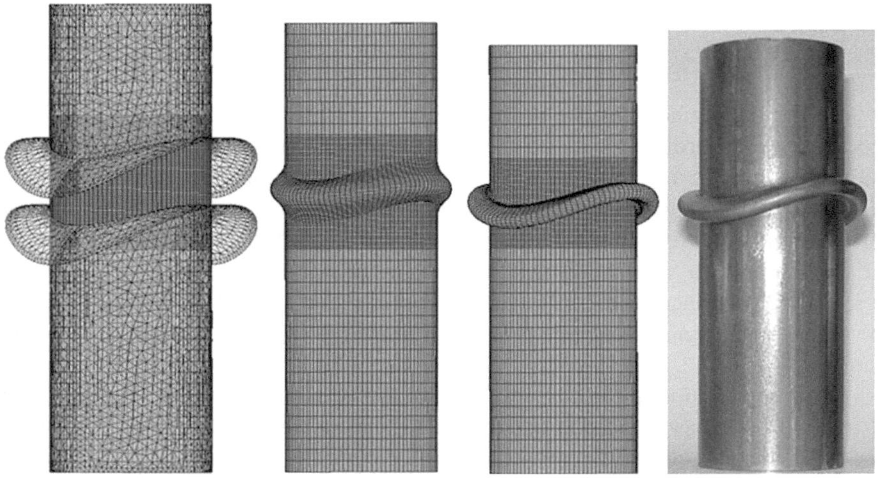

Fig. 12 Finite element model utilized in the numerical simulation of the asymmetric compression beading of thin-walled tubes and photograph of a test sample corresponding to the rightmost numerically predicted geometry of the tube

(e.g. during axisymmetric compression beading), finite element models are built by discretizing the cross section of the tubes and sheets by means of axisymmetric quadrilateral elements and the contour of tools by means of linear elements. In cases of three-dimensional plastic flow resulting from the development of asymmetric compression beading, tubes and sheets are discretized by means of hexahedral elements and tools are discretized by means of triangular spatial elements (Fig. 12).

The numerical simulation of a typical three-dimensional out-of-plane instability wave leading to the formation of an asymmetric compression bead that contains roughly 7,500 elements and 10,500 nodal points (Fig. 12) requires less than 1 h CPU time on a standard laptop computer.

5 Results and Discussion

5.1 Joining Tubes to Sheets

Figure 13 shows the experimental and finite element predicted evolution of the load–displacement curves for clamping a commercial S460MC tube with a reference radius $r_0 = 16$ mm and wall thickness $t_0 = 1.5$ mm to a DC04 (carbon steel) sheet panel with 2.5 mm thickness by a sequence of axisymmetric compression beading followed by tube end flaring (refer to Fig. 7b).

As seen in the figure, numerical and experimental load–displacement curves compare well and the insets, showing details of the finite element simulation at various stages of deformation, provide a good insight on material flow behaviour during the proposed joining process.

The evolution of the load–displacement curve during the first stage of the joining process (compression beading) confirms the development of a plastic instability mode of deformation after approximately 1 mm displacement of the upper die, in close accordance with the critical instability force $F_{cr} \cong 93.5$ kN (refer to Fig. 10).

In what concerns clamping by tube end flaring three different phases may be distinguished: (i) the transient beginning, (ii) the steady-state plastic flow and (iii) the clamping. During the first phase, load increases gradually as the tube is axially compressed and bended against the radiused contour of the upper die. A compression load of approximately 50 kN is obtained at the end of the first stage (refer to 'A' in Fig. 13a) after which the tube starts to bend more pronouncedly in order to match the contour of the die.

The change in material flow produces a slope discontinuity in the load–displacement curve caused by deformation mechanics to evolve into steady-state conditions. The compression load becomes practically constant and at a level considerably below the critical instability load (refer to 'B' in Fig. 13a). The rise of the compression load at the end of the load–displacement curve is related to the

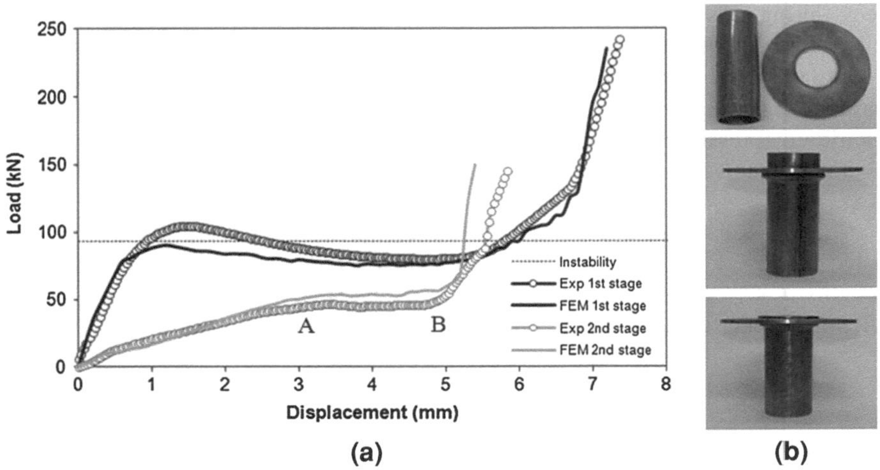

Fig. 13 Joining a tube to a sheet panel by a sequence of axisymmetric compression beading followed by tube end flaring. **a** Experimental and finite element predicted evolution of the load–displacement curves and **b** photograph of the actual joint (before and after being accomplished)

clamp during which the sheet panel is compressed in-between the lower bead and the upper flared tube end.

Figure 13b presents photographs of the actual tube and sheet before and after being joined by the proposed technology and Fig. 14 shows examples of conceptual applications and prototypes developed by the authors. Other examples can be found in the work of Alves et al. [7].

It is worth mentioning the the potential of the proposed joining technology to clamp three sheets made of dissimilar materials (aluminium AA1050-H111, polyvinyl chloride (PVC) and carbon steel DC04) to a commercial S460MC tube. In fact, by varying the type of materials applications can easily extend structural applications and process piping into the fields of thermal and acoustic insulation, among others.

The prototypes of the seat-back bottom frame and hand-brake system for automotives are excellent examples of productivity and environmental advantages of joining tubes to sheets by means of the proposed technology instead of using welding as currently performed in industry. The technical solutions utilized in both prototypes are protected by industrial patents and commercialized by MCG-Mind for Metal.

The proposed technology for joining tubes to sheets by tube forming, at room temperature, can be further enhanced to allow clamping along planes inclined with respect to the tube axis (Fig. 15).

The modification is based on the development of inclined compression beads due to local buckling in thin-walled tubes subjected to axial compression (Fig. 15a). This mode of deformation has never been addressed in the literature

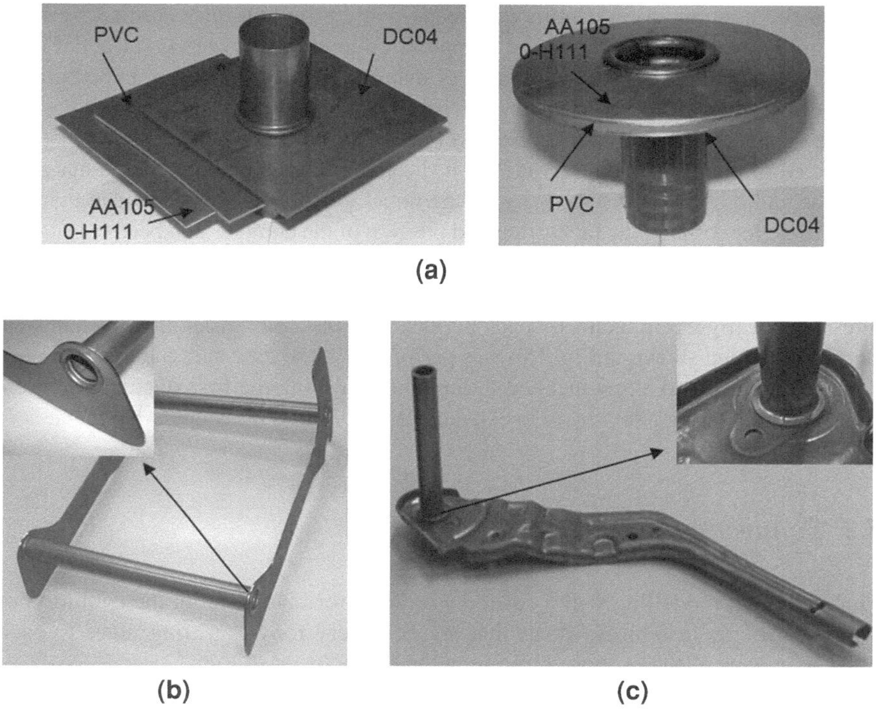

Fig. 14 **a** Joining tubes to sheets of dissimilar materials, **b** prototype of a seat-back bottom frame of an automotive and **c** prototype of a hand-brake system of an automotive

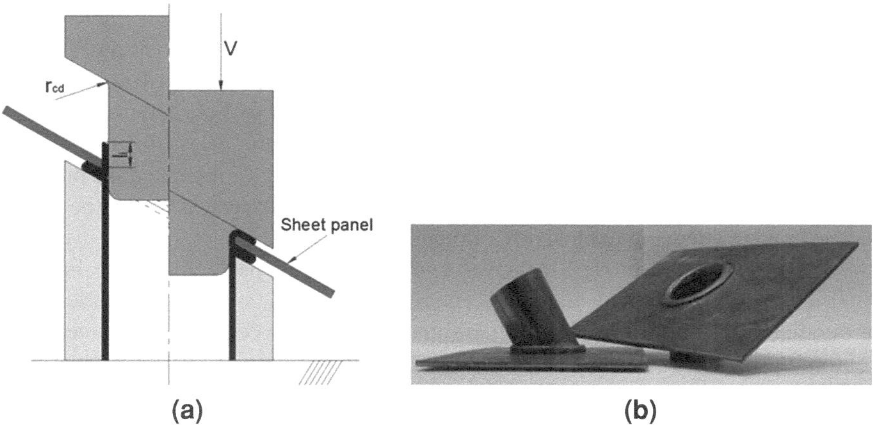

Fig. 15 Modification of the proposed technology to allow joining tubes to sheets along planes inclined with respect to the tube axis. **a** Schematic representation of the tooling system and **b** photograph of an application

and on the contrary to axisymmetric beads, which are naturally formed by local buckling in tubes subjected to axial compression loading between parallel flat dies, inclined beads require the development of instability waves between shaped dies.

The tooling system for producing inclined instability waves in tubes is shown in Fig. 15a, and consist of; (i) two (upper and lower) shaped dies and (ii) an inner mandrel connected to the upper die. The dies are dedicated to a specific reference radius r_0 of the tube and its geometry, together with the initial gap opening l_{gap} between them, is responsible for defining the inclination and position of the compression beads.

Preliminary studies by the authors revealed that slenderness ratios $l_{gap}/r_0 < 0.30$ in commercial S460MC tubes are not feasible because they are below the minimum threshold for triggering inclined beads by means of plasticity instability mechanisms. Because the joint shown in Fig. 15b was performed at the tube end, the upper die has a corner radius r_{cd} to allow inclined flaring after finishing the first stage of compression beading. Flaring allows tube to expand outwards and form a single-lap flange.

5.2 Connecting Tubes

Finite element modelling with i-form was crucial for investigating the evolution of the out-of-plane instability waves that are necessary for connecting tubes to each other by means of the proposed joining technology (refer to Fig. 12).

Finite element modelling also allows understanding the key role played by the internal mandrel in setting up the overall quality of the asymmetric compression beads. In case deformation is performed without a mandrel, the development of the out-of-plane instability waves between contoured dies is the result of a local buckling collapse mechanism that involves inward and outward material flow. The later gives rise to a defect (geometrical depression) on the surface of the tube placed immediately above and below the bead and to non-acceptable variations in the inner diameter of the tube, which are shown in the predicted finite element geometry and photograph of the actual part that are included in Fig. 16a.

The utilization of a mandrel inside the tube not only avoids defects along the surface of the tube but also guarantees the dimension of the inner diameter (which in many applications is a critical dimension) to stay within tolerances. In practical terms, the use of a mandrel eliminates inward material flow and forces the asymmetric beads to develop exclusively outwardly (Fig. 16b).

As experimentally proved by Alves and Martins [8] the width of the beads is significantly influenced by the slenderness ratio l_{gap}/r_0 and can be made larger or smaller by changing the initial gap opening l_{gap} between the upper and lower contoured dies (Fig. 8).

In case of commercial S460MC tubes, if the slenderness ratio is below a minimum threshold (say, $l_{gap}/r_0 < 0.625$) the unsupported free length of the tube is not capable of ensuring local buckling and, therefore, the thickness of the tube will increase uniformly and no compression bead will be formed (refer to the leftmost tubular part in Fig. 6).

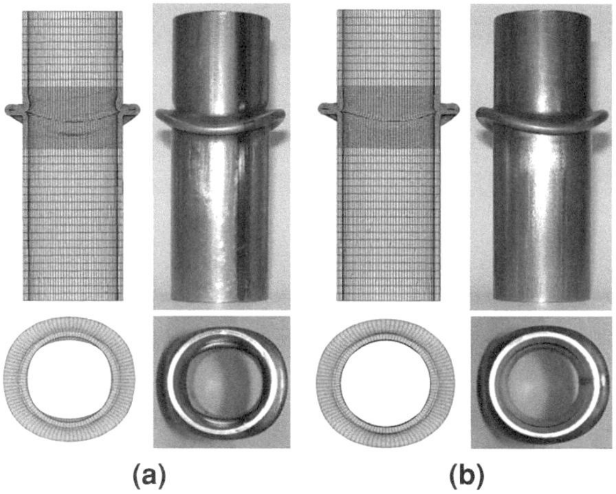

Fig. 16 The role of the internal mandrel on the overall quality of asymmetric compression beads. Finite element predicted geometry and experimental part for a test case performed **a** without a mandrel and **b** with a mandrel

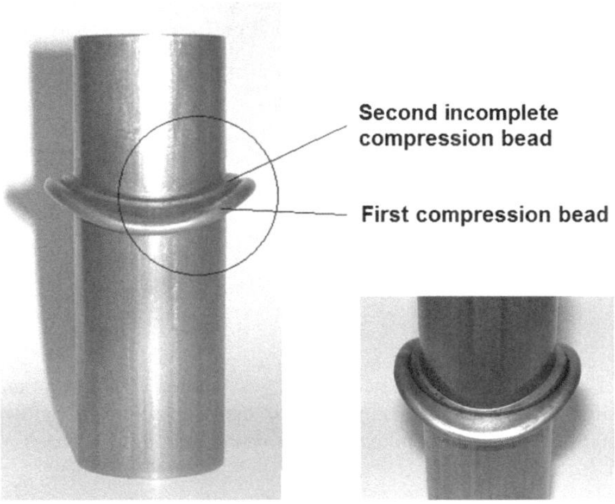

Fig. 17 Development of multiple asymmetric tube beads for an experimental test part with a large value of the slenderness ratio

On the other hand, if the slenderness ratio is large (say, $l_{gap}/r_0 > 1.25$) there is a risk of multiple out-of-plane instability waves interfere with each other as they formed in-between the upper and lower dies. The result of this is shown in Fig. 17, where two asymmetric compression beads are placed on top of each other. As seen in the detail included in Fig. 17, not only the two waves interfere destructively as the second compression bead resulted incomplete due to lack of free gap opening between dies at the end of stroke.

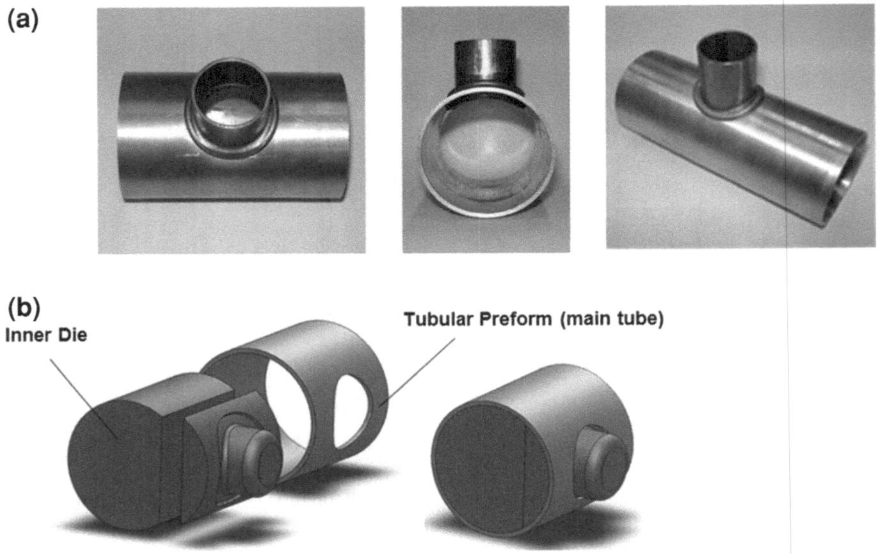

Fig. 18 Tube branching by means of the proposed joining technology. **a** Typical connection and **b** schematic representation of the inner sectioned die and preform of the main tube

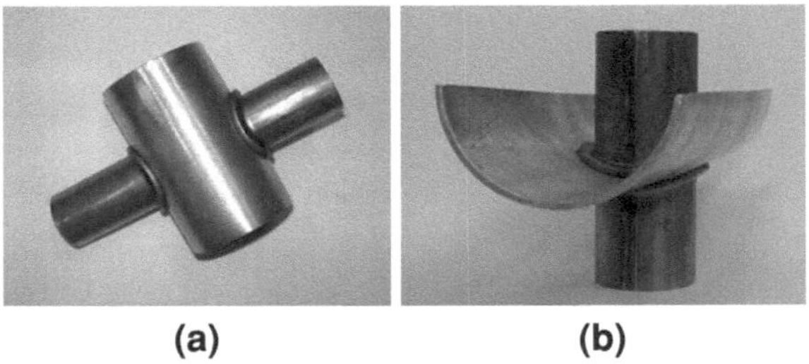

Fig. 19 Examples of application of asymmetric compression beads in tube branching by means of the proposed joining technology. **a** Connection of tubes at nodes in lightweight structural frames and **b** connection between outlet tubes and half round rain gutters

Figure 18 shows a typical connection produced by means of the proposed joining technology without resort to filling materials and heat-cooling cycles that are typical of welding or brazing.

The connection in Fig. 18 is performed through a sequence of two basic steps; (i) firstly, an asymmetric flare is produced at the lower end of the branch tube, making use of an inner sectioned die consisting of two different parts and (ii) secondly, the connection is ensured by means of an asymmetric compression bead to be developed in branch tube along the contour of the pre-cut hole of the main tube.

Figure 19 shows other potential applications of tube joining by means of asymmetric compression beads for producing connection elements (nodes) in industrial and home lightweight structures.

6 Conclusions

This chapter presents a flexible, cost-effective and environmental friendly technology for connecting tubes and joining tubes to sheets that is based on compression beading of thin-walled tubes resulting from plastic instability mechanics. The technology may involve tube end flaring when connections and joints are to be produced at the tube ends and it can successfully eliminate existing technologies based on fasteners, welding and structural adhesive bonding in standard or user made types of connections.

The formability window of the technology is defined by a minimum threshold of the slenderness ratio below which it is not possible to trigger local buckling and a maximum threshold above which there is a risk of multiple instability waves to develop and interfere with each other. The utilization of inner mandrels is important for avoiding defects and variations in dimensions in the regions of the tube that are placed immediately above and below the compression beads.

The possibility of connecting tubes of different materials and joining metallic sheets with polymeric tubes, or vice versa, offers potential to foster innovation and creativity in tube joining, to develop new products and to open new markets in automotive, process piping and assembly of lightweight frame structures, among other industries.

Acknowledgments The authors would like to acknowledge the opportunity and support provided by the automotive division of MCG—Mind for Metal, Carregado, Portugal.
The work of Eduardo Dias, Nuno Rodrigues and Bernardo Calado is also greatly appreciated.

References

1. Alexander JM (1960) An approximate analysis of the collapse of thin cylindrical shells under axial loading. Q J Mech Appl Mech 13:10–15
2. Allan T (1968) Investigation of the behaviour of cylindrical tubes subject to axial compressive forces. J Mech Eng Sci 10:182–197
3. Rosa PAR, Rodrigues JMC, Martins PAF (2003) External inversion of thin-walled tubes using a die: experimental and theoretical investigation. Int J Mach Tools Manuf 43:787–796
4. Gouveia BPP, Alves ML, Rosa PAR, Martins PAF (2006) Compression beading and nosing of thin-walled tubes using a die: experimental and theoretical investigation. Int J Mech Mater Des 3:7–16
5. Al-Qureshi HA, Das MK (1976) Thin-walled tube cropping using elastomer. Int J Mach Tool Des Res 16:77–85
6. Miller G (2003) Tube forming processes: a comprehensive guide. SME-Society of manufacturing Engineers, Dearborn
7. Alves LM, Dias EJ, Martins PAF (2011) Joining sheet panels to thin-walled tubular profiles by tube end forming. J Clean Prod 19:712–719
8. Alves LM, Martins PAF (2012) Tube branching by means of asymmetric compression beading. J Mater Process Technol 212:1200–1208
9. Shanley FR (1947) Inelastic column theory. J Aeronaut Sci 14:261–268
10. Timoshenko S (1940) Theory of plates and shells. McGraw-Hill, New York
11. Alves LM, Nielsen CV, Martins PAF (2011) Revisiting the fundamentals and capabilities of the stack compression test. Exp Mech 51:1565–1572

12. Kobayashi S, Oh SI, Altan T (1989) Metal forming and the finite element method. Oxford University Press, New York
13. Alves ML, Rodrigues JMC, Martins PAF (2003) Simulation of three-dimensional bulk forming processes by the finite element flow formulation, modelling and simulation in materials science and engineering. Inst Phys 11:803–821

Concepts, Methods, and Strategies for Zero-Waste in Manufacturing

Matthew J. Franchetti and Alexander Spivak

Abstract Zero waste in manufacturing is becoming very prevalent in the United States. An effective way of economically achieving zero waste is through the waste assessment process and systems approach. This approach allows businesses to establish green team and process, complete the assessment, determine waste reduction opportunities and economically and efficiently implement them. The system helps to establish an effective way to continuously work on waste reduction, compare alternatives and periodically validate results to achieve stated goals. There are number of strategies and techniques that businesses may use to achieve a zero waste goal. These strategies include techniques for reduction of paper products, packaging, containers and food, as well as identification of other recyclable products. Strategies for collection stations, source reduction, transportation and delivery, alternative uses of products, "green" purchasing and HR actions are also presented in the chapter. It is author's hope that readers will find new methods and approaches in this chapter that they will implement at their facilities.

1 Concepts of Zero Waste in Manufacturing

Industrial solid waste management and minimization is an effective tool that is a necessary aspect in business today. Firms are discovering it to be a profitable approach, while at the same time enhancing their company image as environmentally conscious and responsible. In the extremely competitive world of manufacturing, waste disposal is often overlooked. There are measurable costs associated with excess raw material, scrap parts, insufficient use of resources,

M. J. Franchetti (✉) · A. Spivak
Department of Mechanical, Industrial, and Manufacturing Engineering,
University of Toledo, Toledo OH 43606, USA
e-mail: matthew.franchetti@utoledo.edu

A. Spivak
e-mail: alexander.spivak@utoledo.edu

J. P. Davim (ed.), *Green Manufacturing Processes and Systems*, Materials Forming, Machining and Tribology, DOI: 10.1007/978-3-642-33792-5_4,

and outdated materials. All these factors contribute to a company's waste stream, and must be included when calculating cost of disposal. The concept of zero landfill facilities has also become very prevalent in the US. Many organizations have committed to preventing their waste streams from entering landfills through reduction, reuse, recycling, composting and energy conversion. However, many companies do not possess the proper resources to effectively and efficiently manage and minimize their solid waste steams. External consultants are often required to enable companies to optimize their solid waste handling and disposal and minimize waste costs. Waste analysis and minimization can be advantageous to companies in improving their financial position and competitiveness.

1.1 Terms and Definitions

Most terms and definitions were obtained from State of Connecticut official website (www.ct.gov).

Air Pollution—The presence in the outdoor atmosphere of one or more air pollutants or any combination thereof in such quantities and of such characteristics and duration as to be, or be likely to be, injurious to public welfare, to the health of human, plant or animal life, or to property, or as unreasonably to interfere with the enjoyment of life and property.

Best Management Practice (BMP)—Methods that have been determined to be the most effective, practical means of preventing or reducing pollution from non-point sources. (EPA Glossary)

Biodegradable—Capable of decomposing under natural conditions. (EPA Glossary)

Biodegradable Plastic—A degradable plastic in which the degradation results from the action of naturally occurring microorganisms such as bacteria, fungi, and algae. (ASTM Standard Specification D6400)

Carbon Footprint—A measure of the impact our activities have on the environment, and in particular climate change. It relates to the amount of greenhouse gases produced in our day-to-day lives through burning fossil fuels for electricity, heating and transportation etc. The carbon footprint is a measurement of all greenhouse gases we individually produce and has units of tons (or kg) of carbon dioxide equivalent. (Carbonfootprint.com)

Commingled recyclables—Mixed recyclables that are collected together (Business Recycling Manual, INFORM & Recourse Systems, 1991)

Compact Fluorescent Lamp (CFL)—Small fluorescent lamps used as more efficient alternatives to incandescent lighting. Also called PL, CFL, Twin-Tube, or BIAX lamps. (EPA Glossary)

Compaction—Reduction of the bulk of solid waste by rolling and tamping. (EPA Glossary)

Compost—A humus or soil-like material created from aerobic, microbial decomposition of organic materials such as food scraps, yard trimmings, and manure (EPA Glossary)

Compostable Plastic—A plastic that undergoes degradation by biological processes during composting to yield CO2, water, inorganic compounds, and biomass at a rate consistent with other known compostable materials and leave no visible, distinguishable or toxic residue. (ASTM Standard Specification D6400)

Composting—A process of accelerated biological decomposition of organic material under controlled conditions. (CGS Section 22a-207a (1))

Corrugated cardboard—Cardboard with corrugations (can be glued to flat cardboard on one or both sides) [syn: corrugated board] (WordNet® 3.0, 2006, Princeton University via Dictionary.com)

Degradable Plastic—A plastic designed to undergo a significant change in its chemical structure under specific environmental conditions, resulting in a loss of some properties that may be measured by standard test methods appropriate to the plastic and the application in a period of time that determines its classification. (ASTM Standard Specification D6400)

Energy Recovery—Obtaining energy from waste through a variety of processes (e.g. combustion). (EPA Glossary)

Food Waste—Uneaten food and food preparation wastes from residences and commercial establishments such as grocery stores, restaurants, and produce stands, institutional cafeterias and kitchens, and industrial sources like employee lunchrooms. (EPA Glossary)

Grasscycling—Source reduction activities in which grass clippings are left on the lawn after mowing. (EPA Glossary)

High Density Polyethylene (HDPE)—A type of plastic used to make milk jugs and other rigid plastic bottles (City Cycle, 1990) A material used to make plastic bottles and other products that produces toxic fumes when burned. (EPA Glossary) Coded #2

Materials Recovery Facility (MRF)—A facility that processes residentially collected mixed recyclables into new products available for market. (EPA Glossary)

Office paper—Used or discarded high grade white paper and manilla paper, including paper utilized for file folders, tab cards, writing, typing, printing, computer printing, and photocopying, which is suitable for recycling and which has a minimum of contamination (City Cycle)

Pallet—A wooden platform used in connection with a forklift for moving bales or other large items. Also called a 'skid'. (Business Recycling Manual, INFORM & Recourse Systems, 1991)

Polyethylene Terephthalate (PETE)—A type of plastic used to make soft drink bottles and other kinds of food containers. PET is also used to make fabric. Coded #1. (Earth911.org)

Polyvinyl Chloride (PVC)—A tough, environmentally indestructible plastic that releases hydrochloric acid when burned. (EPA Glossary)

Recycled-content—Products are made from materials that would otherwise have been discarded. That means these products are made totally or partially from material contained in the products you recycle, like aluminum soda cans or newspaper. Recycled-content products also can be items that are rebuilt or re-manufactured from used products such as toner cartridges or computers. (Earth 911)

Reuse—Using a product or component of municipal solid waste in its original form more than once; e.g., refilling a glass bottle that has been returned or using a coffee can to hold nuts and bolts. (EPA Glossary)

Solid Waste—Unwanted or discarded solid, liquid, semisolid or contained gaseous material, including, but not limited to, demolition debris, material burned or otherwise processed at a resources recovery facility or incinerator, material processed at a recycling facility and sludges or other residue from a water pollution abatement facility, water supply treatment plant or air pollution control facility. (CGS Section 22a-207 (3))

Source Reduction—Reducing the amount of materials entering the waste stream from a specific source by redesigning products or patterns of production or consumption (e.g., using returnable beverage containers). Synonymous with waste reduction. (EPA Glossary)

Sustainability—meeting the needs of the present without compromising the ability of future generations to meet their own needs. (EPA Glossary)

Toxic Waste—A waste that can produce injury if inhaled, swallowed, or absorbed through the skin. (EPA Glossary)

Used Oil—(1) Spent motor oil from passenger cars and trucks collected at specified locations for recycling (EPA Glossary) (2) Fried and baked corn, olive, sunflower, and other edible oil—often may be used for biodiesel production.

Waste-to-Energy Facility/Municipal-Waste Combustor—Facility where recovered municipal solid waste is converted into a usable form of energy, usually via combustion. (EPA Glossary)

Zero Waste—Zero Waste is a goal that is both pragmatic and visionary, to guide people to emulate sustainable natural cycles, where all discarded materials are resources for others to use. Zero Waste means designing and managing products and processes to reduce the volume and toxicity of waste and materials, conserve and recover all resources, and not burn or bury them. Implementing Zero Waste will eliminate all discharges to land, water or air that may be a threat to planetary, human, animal or plant health. (Zero Waste International Alliance)

2 Waste Assessment Process and Systems Approach

After the need has been identified to minimize solid waste generation and top management supports and allocates resources to the effort, a solid waste assessment can be conducted. The solid waste assessment is one of the most important steps of the solid waste minimization process because the data generated provides the team and management with a much greater understanding of the types and amounts of waste generated by the company. These data can be invaluable in the design and implementation of a waste reduction program. Before ‘rushing in’ to conduct a solid waste assessment, proper planning should be done to ensure the scope, goal, and timeline of the project fits into the strategic plan of the organization. This chapter provides the systems approach framework to ensure that these goals are met using the five-step Six Sigma DMAIC process:

- Define—establish the project team, determine the goal, and allocate the necessary resources
- Measure—conduct a solid waste assessment and review existing records
- Analyze—statistically analyze the data collected to identify trends
- Improve—modify the process to meet the organizational goals established in Define stage
- Control—ensure that the improvement initiatives stay in place with continuous improvement and feedback

The systems approach examines the organization as a whole, or a sum of all business processes to achieve established goals. The concept is to use data to develop comprehensive system wide changes that will drive environmental and economic performance versus routine incremental improvements. In addition, several examples will be provided to further explore and explain each step of the framework. The solid waste minimization process consists of 11 steps:

1. Establish the solid waste minimization team and charter
2. Review existing solid waste and recycling records
3. Create process flow charts and conduct throughput analyses
4. Conduct the solid waste sorts at the facility
5. Analyze the data to determine annual generation by work unit or area and establish baseline data
6. Identify major waste minimization opportunities
7. Determine, evaluate, and select waste minimization process, equipment, and method improvement alternatives
8. Develop the waste minimization deployment and execution plan
9. Execute and implement the waste minimization plan and timeline
10. Validate the program versus goals
11. Monitor and continually improve performance

2.1 Step 1: Establish the Team and Define the Project

After identifying the need to minimize solid waste and gaining top management support, the first step in process involves establishing the team and defining goals. The key outcomes and deliverables of this step are:

- A letter of support from top management to all employees
- Team leader and member identification
- Initial training for the team

- Project goal identification including metrics
- Team charter
- Project timeline
- Project budget

The goals of the project should be clearly expressed as soon as possible. This will provide the team with much needed direction and serve as the gauge to evaluate all team activities and accomplishments. The goals should provide a specific direction for the project and not vague generalized improvement slogan such as "to become an environmental leader in the automobile engine manufacturing field" or "to reduce the organization's carbon footprint". A more specific, SMART goal, such as:

- Reduce the amount of solid waste generated per year by 5 % from the baseline year of 2009.
- Increase the recycling rate for metals and paper products by 10 % from the baseline year of 2009.
- Utilize environmental improvements as a strategic weapon to provide a cost benefit of 10 % versus the baseline year of 2009 for expenditures and revenues solid waste removal and recycling efforts.

The project timeline is one of the key deliverables from the 'define' stage. The timeline tracks project performance versus established goals and serves as the strategic implementation plan. The timeline should be viewed as a control document to evaluate the progress of the team versus pre-established milestones. Proper planning is needed to ensure the timeline is achievable and will meet the goals of the project. Below are some general guidelines for the time required for each step solid waste minimization process.

Please note that the timeline assumes that the team will be devoting approximately 30 % of their time to the project and 70 % of their time to normal job duties. On average a successful project requires approximately three to four months to complete. Below is the timeline that was created for the case study (Table 1).

2.2 Step 2: Existing Record Review

Reviewing the existing organizational records for solid waste and recycling usually provides significant insight into the amounts, types, and patterns of waste generation. Very useful data and information can be gained to help focus the efforts of the team and eliminate the need for the collection of existing raw data. Collecting high quality records will save a great deal of time, money and effort versus raw data collection via a trash sort or facility walk-through. The types of records to collect include:

- Purchasing, inventory, maintenance, and operating logs
- Supply, equipment, and raw material invoices

Table 1 Project Timeline

Solid Waste Minimization Step:	Time (weeks)
1. Establish the solid waste minimization team and charter	2
2. Review existing solid waste and recycling records	2
3. Create process flow charts and conduct throughput analyses	2
4. Conduct the solid waste sorts at the facility	2
5. Analyze the data to determine annual generation by work unit or area	2
6. Establish baseline data	1
7. Identify waste minimization opportunities	6
8. Develop process and method improvements alternatives to minimize solid waste	6
9. Determine vendors and service providers to assist in waste minimization	3
10. Compare and decide among alternatives	1
11. Develop the waste minimization deployment and execution plan	2
12. Execute and implement the waste minimization plan and timeline	4
13. Validate the process versus goals	1
14. Monitor and continually improve performance	1
TOTAL	35

- Equipment service contracts
- Repair invoices
- Waste hauling and disposal records and contracts (including one year of amounts and fees collected)
- Contracts with recycling facilities and records of earned revenues from recycling.
- Major Equipment List
- Production Schedule (representative of a year)
- Company Brochure or Product Information
- Material Safety Data Sheets (MSDS)
- Facility Layout (hard copy plus CAD copy, if available)
- Process Flow Diagrams

The purpose of reviewing these records is to determine total amount of waste generated annually, the total amount of material recycled annually, and the total waste removal and recycling cost structure. A review of purchasing records can also be beneficial to 'backtrack' into an estimate of waste components generated. For example, shipments often arrive in cardboard containers. By researching the number of containers received per year and estimating the weight per carton, the team could estimate the total weight of cardboard boxes disposed each year. Often times, the janitorial staff, maintenance, purchasing, and accounting staff will be most useful when gathering these records. In addition, customer service at the waste hauling and recycling companies may record more detailed records than the information that appears on monthly invoices.

2.3 Step 3: Process Mapping and Production Analysis

The goal of this step is to aid the team in fully understanding the business processes and capabilities of the facility. An understanding of these processes is crucial in developing alternatives to reduce solid waste. A process flow chart is a hierarchical method for displaying processes that illustrates how a product or transaction is processed. It is a visual representation of the work-flow either within a process or an image of the entire operation. Process mapping comprises a stream of activities that transforms a well defined input or set of inputs into a pre-defined set of outputs.

A well developed process flow chart or map should allow people unfamiliar with the process to understand the interaction of causes during the work-flow and contain additional information relating to the solid waste minimization project (such as tons of waste generated per year and annual cost of disposal).

2.4 Step 4: Solid Waste Sorts

Solid waste sorts provide detailed data regarding the composition of an organization's waste stream. Via the data collected from the solid waste sorts, the organization's waste stream can be characterized into the various materials that comprise the entire stream, including the annual amounts generated and the percentage that each component contributes to the entire waste stream. This data is invaluable when evaluating cost effective methods or process changes to divert these components from landfills. The waste sort itself is affectionately referred to as 'dumpster diving', since the team will physically collect, sort, and weigh a representative sample of the organization's waste. This remainder of this section discusses the process to conduct a waste sort, including the required preparation, tools required, a step-by-step guide, and also provides data collection forms.

The team meeting prior the waste sort should be held to get the team on the same page. The primary outcomes from this meeting are training and a waste sort plan. The training should be conducted by the technical expert and focus on the use of tools and data collection form. The waste sort plan should assign team members to the various areas of the facility to conduct data collection effectively and efficiently. The equipment required for the waste sort includes:

- Gloves
- Yard sticks
- Plastic bags
- Scales
- Clipboards (with the data collection from)

Based on the size of the facility, it may be necessary to recruit additional support to collect data during the waste audit. If additional help is used, these individuals should receive the same training as the core team. An approach that seems to work well is to assign one temporary team member from each work unit within the facility. For example, assign a supervisor or shop floor worker from each production work area such as the metal stamping unit, the paint shop, and the accounting offices (work units will differ by business type). The advantages of assigning temporary team members within each work unit include faster and more accurate data collection. As a member of the work unit in which data is being collected, the temporary team member will possess specialized knowledge on the waste generation types and amounts.

The data collection form is the most important document of the waste sort. The data collected with the form will be used to extrapolate the annual generation for the facility so care should be taken when collecting the data to ensure accuracy. At a minimum, the required information on the data collection form is:

- The date the data was collected
- The team members collecting the data (this is very useful if follow up or clarification is needed)
- Work unit and location of the waste receptacle
- Source of the waste (previous operation or supplier)
- Disposal method (baler, compactor, recycler, un-compacted dumpster)
- Size of the container in cubic feet (can be derived from length, width and height measurements)
- The container type (desk side, recycling bin, dumpster)
- Percent full
- Times emptied (per day, week or month)
- Container contents and percent of each component
- Condition of material (loose, compacted, baled)
- Notes and comments that may be useful when analyzing the data (including names and contact information for work units members with specialized knowledge of the waste or generation levels)

2.5 Step 5: Data Analysis

The primary outcome of the analysis phase is the annualized waste generation baseline data for the facility. This baseline data should be broken down by the component waste stream (paper, metal, etc.), work unit of generation, and how the component waste stream is currently handled (landfill, recycled, burned, etc.).

Each component should be given in terms of weight (tons per year) and volume (cubic yards per year). The key questions that are answered from this analysis are:

- What are the waste streams generated from the facility and how much?
- Which processes of operations do these waste streams come from?
- Which waste streams are classified as hazardous and which are not? What makes them hazardous?
- What are the input materials used that generates the waste streams of a particular process or facility area?
- How much of a particular input material enters each waste stream?
- How much of a raw material can be accounted for through fugitive losses?
- How efficient is the process?
- Are any unnecessary wastes generated by mixing otherwise recyclable hazardous materials with other process wastes?
- What type of housekeeping practices are used to limit the quantity of wastes generated?
- What types of process controls are used to improve process efficiency?
- How much money is the company paying to dispose of solid waste and how much revenue does it generate from the sale of recyclable materials?

To answer these questions and to generate the baseline data, the existing records collected, the data gathered during the waste audit, and team member knowledge will be used. Additional data collection or verification may be required during the analysis portion.

After all data has been entered into the spreadsheet, annual generation in terms of weight and volume can be tabulated. To estimate the annual waste stream in both terms of weight and volume is very important because, in general, waste haulers charge based on volume (cubic yards that fill a dumpster) and processors pay for recyclable materials based on weight (tons in a bale). Below is an example calculation for cardboard (OCC).

A company reported using a dumpster of 12 cubic yards that was used exclusively for compacted OCC (cardboard) that was emptied two times per month by a recycling vendor. This data was converted into annual tonnage using the following equation:

$$Tons\ per\ Year = (Dumpster\ Size\ in\ cubic\ yards) * (Times\ Emptied\ per\ Month) * (12\ Monthsper\ Year) * (EPA\ Average\ Material\ Density - tons/cubic\ yards)$$

$$OCC = (12\ cubic\ yards) * (2/month) * (12 months/year) * (0.45\ tons/cubic\ yard) = 129.6\ tons\ of\ OCC\ per\ year$$

2.6 Step 6: Identify Major Waste Minimization Opportunities

After the baseline data has been calculated, the assessment team can begin to investigate individual components in the waste stream that should be targeted for reduction, reuse, or recycling. A useful method to accomplish this task is to conduct a Pareto analysis. A Pareto analysis is a statistical technique in decision making that is used for the selection of a limited number of tasks that produce the significant overall effect. Pareto analysis is a formal technique useful where many possible courses of action are competing for your attention. In essence, the problem-solver estimates the benefit delivered by each action, then selects a number of the most effective actions that deliver a total benefit reasonably close to the maximal possible one. The analysis uses the Pareto principle—a large majority of effects, in this case waste generation are produced by a few key causes, in this case waste components. The Pareto principle is also known as the 80/20 rule that 80 % of the effects are caused by 20 % of the causes. The idea is to identify the 20 % significant wastes components that generation 80 % if the total waste and then target the 20 % significant causes for waste minimization.

2.7 Step 7: Determine, Evaluate, and Select Waste Minimization Alternatives

Once the major waste streams have been quantified, the team can begin to develop alternatives to minimize the solid waste and move closer to the ultimate goal of the project. In this phase of the project, the team identifies alternatives to minimize the major components of the waste stream and evaluates the economic and operational feasibility while rating each alternative on its ability to achieve the waste minimization goal. This section covers the process to generate, screen, and select waste minimization alternatives. In addition, a comprehensive list of common materials that can be reduced, reused or recycled and a list of common waste minimization alternatives is also provided.

2.7.1 Generating Alternatives

The alternatives are based on the existing records review, the waste audit results, and the analysis phase. Various methods and tools are available to develop the initial list of alternatives. The environment in which these alternatives are created should be done in one that encourages creativity and free thinking by the team. Following is a suggested list of methods to identify and create these alternatives:

- Discussions with trade associations
- Discussions with plant engineers and operators
- Internet and literature reviews
- Information available from Federal, state or local governments
- Discussions with equipment manufacturers or vendors
- Discussions with environmental or business consultants
- Brainstorming
- Benchmarking

Common Waste Minimization Alternatives (Materials and Methods)

This section provides a brief list of common solid waste minimization alternatives that many companies have successfully implemented. When considering alternatives to minimize waste, the solid waste minimization hierarchy should be considered. Consideration should first be given to reducing the waste (find a process or purchasing change to prevent generating the waste in the first place), next reuse the waste item, and next recycle the waste item, and finally disposal at the landfill. Source reduction can be accomplished through process modifications, technology changes, input material changes, or product changes. The following list provides commonly applied waste reduction and reuse opportunities that many companies have successfully implemented. Most of these are relatively low cost and are considered the 'low hanging fruit', or simple to launch.

- Office paper—many easy options exist to reduce office paper usage including implementing an organization wide double-sided copying policy (set the defaults of copiers and printers to print double-sided), reuse old paper as scratch paper, put company bulletins in electronic form (email), centralize files to reduce redundant copies, save files electronically versus hard copy, and donate old magazines to hospitals or other organizations.
- Packaging—order merchandise in bulk, work with suppliers to minimize packing materials, establish a reuse policy for cardboard boxes, implement returnable containers, reuse shredded newspaper as packing material.
- Equipment—use rechargeable batteries, reuse old tires for landscaping or pavement, install reusable filters, donate old furniture, sell obsolete equipment and computers.
- Organic waste—compost yard trimmings, choose low maintenance landscape designs, use mulching mowers.
- Inventory/purchasing—set up an area in the facility where employees can exchange used items, purchase more durable products, order in bulk to reduce packaging supplies, use a waste exchange program.

2.7.2 Screening Alternatives

The process of creating waste minimization alternatives can generate hundreds of options. It would be very time consuming for the team to conduct detailed financial and operational feasibility evaluations on each option. A quick screening process can help to quickly identify the options worthy of full evaluations and possible inclusion in the waste minimization program. Additionally, non effective options can be weeded out, saving the team time and money in the evaluation process. An effective screening process should be based on the original goals of the project and at a minimum should examine:

- Expected solid waste reduction (tons per year)
- Expected start up costs

- Impact to waste removal costs ($ per year)
- Impact to purchasing costs ($ per year)
- Impact on employee moral
- Ease of implementation

2.7.3 Analyzing and Selecting Alternatives

After trimming down the list of alternatives via the screening process, the remaining alternatives should be further analyzed to determine the best fit for the organization to minimize solid waste and hence include in the program. The analysis process focuses on identifying the benefits, costs, and drawbacks of each alternative. To accomplish this, each alternative is evaluated based on:

- Impact on the program goal
- Technical feasibility
- Operational feasibility
- Economic feasibility
- Sustainability
- Organizational culture feasibility

Technical and operational feasibility is concerned with whether the proper resources exist or are reasonably attainable to implement a specific alternative. This includes the square footage of the building, existing and available utilities, existing processing and material handling equipment, quality requirements, and skill level of employees. During this process, product specifications and facility constraints should be taken into account. Typical technical evaluation criteria includes:

- Available space in the facility
- Safety
- Compatibility with current work processes and material handling
- Impact on product quality
- Required technologies and utilities (power, compressed air, data links)
- Knowledge and skills required to operating and maintain the alternative
- Addition labor requirements
- Impact on product marketing
- Implementation time

When evaluating technical feasibility, the facility engineers or consultants should be contacted for input. In addition it is also wise to discuss the technical aspects with workers directly impacted by the change such as production and maintenance. If an alternative calls for a change in raw materials, the effect on the quality the final project must be evaluated. If an alternative does not meet the technical requirements of the organization, it should be removed from consideration. From a technical standpoint, the three areas that require additional evaluation are:

- Equipment modifications or purchases
- Process changes
- Material changes

From an economic standpoint, traditional financial evaluation is the most effective method to analyze alternatives. These measures include the payback period (discounted payback period), internal rate of return, and net present value for each alternative. If the organization has a standard financial evaluation process, this should be completed for each alternative. The accounting or finance department would have this information. To perform these financial analyses, revenue and cost data must be gathered and should be based on the expectations for the alternatives. This is more complicated that is sounds, especially if a project will have an impact on the number the required labor hours, utility costs, and productivity, not to mention initial investments. A comprehensive estimation of the cost impacts (revenues and costs) per year over the life of the alternative is required to begin the analysis. The first step of the economic evaluation process is to determine these costs. These costs include capital costs (or initial investment), operating costs/savings, operating revenue, and salvage values for each waste minimization alternative.

2.8 Step 8: Documentation and Development of the Deployment Plan

The goal of this phase of the waste minimization process is to translate the list of accepted alternatives into an achievable implementation plan and to document the selected alternatives. The remainder of this section provides an overview of the deployment plan, a discussion regarding obtaining funding, and details regarding each of the twelve sections of the deployment plan. Finally, a template for the deployment plan is provided.

2.9 Step 9: Implementation and Execution of the Solid Waste Minimization Plan

A well developed deployment plan based on viable options will yield poor results if the plan is not executed properly. There is no such thing as "over-communication" when it comes to rolling out a new project or program. The three key components of a successful implementation and execution are following the deployment plan, communication, and recognizing the need to adjust in certain circumstances.

To facilitate the communication process, at a minimum, weekly progress meetings should be held with all key stakeholders. These meetings should focus on the status of each project versus the timeline and established goals. An agenda and the project timeline should be prepared in advanced and serve to lead the discussion. The task leader (as determined in the deployment plan) should take the lead role in discussing the status of each project or program. Any obstacles or delays should be discussed so that the team may determine solutions.

During the deployment process it is critical not to overwhelm employees with process changes. Effort should be taken to ensure that all employees are aware of upcoming changes, timelines, and the reasons behind the change. This can be accomplished with service talks, postings or newsletters in paychecks. All three options may be used to ensure that the message is heard and that employees are not confused and buy into the programs.

In general, less effort is required for operational and process changes. These options can usually be implemented in a much quicker fashion than equipment or material changes. A general outline of the scope of an implementation effort is provided below:

- Approve the project or program
- Finalize the specifications and design for each alternative
- Submit and gather bid requests and quotes (if necessary)
- Complete and submit a purchase order
- Receive and install the equipment
- Finalize operating and maintenance procedures
- Train affected employees
- Start the project or program
- Complete regulatory inspections
- Track implemented project cost savings and waste reductions

2.10 Step 10: Validate the Program Versus Goals

Many companies require a validation process to ensure that projects and programs have met the goals that were set at the onset of the project. This includes validating the project or program was installed at or below cost, that it is operating within the expense and revenue limits, and that it is achieving the waste reduction goals. Even if an organization does not require a validation process, it can be a very valuable tool for future planning processes to identify were estimation errors occurred and take effort to correct them. Alternatives that do not meet the established goals or expected performance expectations may require rework or modifications. It is also critical to store warranties and contracts from vendors prior to the installation of the equipment. Also, the experience gained in implementing an option at one facility can be used to reduce the problems and costs of implementing options at subsequent facilities.

An alternative performance analysis should be completed for each equipment, process, or material change. The analysis provides a standardized method to compare project performance again estimates in terms of:

- Project duration
- Implementation cost
- Operating expenses and revenue
- Waste reduction volume
- Cycle time and productivity
- Product or process quality
- Safety

2.11 Step 11: Monitor and Continually Improve Performance

After the waste minimization program has been implemented and validated it must be monitored on a periodic basis to ensure that it is still performing as planned and to make any necessary adjustments. This includes monitoring the waste reduction amounts, operational and financial performance versus the goals. In addition, emphasis should be placed on continuous improvement to enhance current waste reduction programs and to identify new opportunities. It may be beneficial to conduct period waste assessments, facility walk-through, or employee interviews by the original waste reduction team to accomplish these goals. When evaluation the program it is important to:

- Keep track of program success and to build on past successes.
- Identify new ideas for waste reduction
- Identify areas needing improvement
- Document compliance with state or local regulations
- Determine the effect of new additions to the facility or program
- Keep employees informed and motivated

In addition, considering reviewing the organization's waste removal receipts and purchasing records on at least a quarterly basis to ensure that the waste minimization program is working. New product or process changes should also be evaluated at the onset to ensure that the design minimizes environmental impact. This is easily accomplished by adding "waste minimization review" to the new product or process checklist or standard operating procedure.

The waste minimization program is a continuing versus a one time project. Generally, the first waste assessment and implemented alternatives will target only the high volume waste streams. Once these high volume streams have been reduced, the team and focus on smaller volume waste streams. From a systems standpoint, the ultimate goal of the team is to minimize all input materials into the facility and by-products generated by the facility. The frequency in which the additional waste assessments are conducted will depend on the budget of the company. In general, organizations that conduct assessments one to four times per year have achieved paybacks. In addition, if there special circumstances that indicate the need for further review a waste assessment should be conducted, these special circumstances include:

- A change in raw material or product requirements
- Higher waste management costs
- New regulations
- New technology
- A major event with undesirable environmental consequences (such as a major spill)

To be truly effective, an organizational culture of waste minimization must be fostered within the organization. Executive management must ensure this through repeated communications and acknowledgements for success stories from individuals or business units. This will make waste minimization an integral part of the organization's operations.

3 Common Strategies for Zero Waste

The goal of Zero Waste is attainable through strategic reduction of the solid waste. Along with "low hanging fruit" reductions, such as paper products, aluminum cans, old corrugated containers recycling and source reduction, there are many other techniques that are available. These techniques include recyclable product identification, paper product management, collection stations, packaging management, source reduction applications, food management, alternative uses for products, "green" purchases and "green" partnerships, delivery and transportation measures, and HR/PR measures. The presented list of techniques is not exhaustive, but rather a source of ideas.

3.1 Recyclable Product Identification

Prior to the start of the recycling program, it is important to understand which bi-products (not trash) can be recycled or reused. Such products may be identified through assessment. The assessment may be conducted internally, through county or city Solid Waste Management District (free of charge), through other grants (typically free of charge) or by hiring for-profit assessment company. Assessment will include product identification as well as recycling methods. Almost all bi-products may be recycled, reused, reduced at the source or composted. Therefore, it is very possible to achieve a goal of zero landfill. Typically, net profit is generated when zero-landfill goal is achieved. The list of some products that may be reused or recycled is:

ABS plastic, alcohol/flux waste, aluminum, aluminum cans, amber glass, antifreeze, appliances, asphalt, bailing wire (compressed), batteries (lead-acid), batteries (lead-free), binders, blue wrap from operating rooms, bottles (blue glass) bottles (brown glass), bottles (green glass), bottles (plastic), bottles

(transparent glass), boxes/packaging, brass, bubble wrap, buckets, building and demolition materials, cardboard (OCC), carpet backing, carpets, CD disks, CD players, cell phones, ceramics, CFL bulbs, circuit board scrap, circuit boards, computer paper, computer scrap, computers, concrete, construction material, cooking oil, copper, CPVC plumbing, CRTs, demolition debris, drywall, DVDs, electrical equipment, e-waste, film, fluid bags and coverings, fluorescent tubes, foam, furniture, glass, glassware, GPPS (black trays from printer assembly, grease from kitchen, gypsum, HIPS (black, random and mixed polystyrene), IC tubes, industrial plastics, ink, ink cartridges, ink sludge from ink treatment, ink treatment resin filters, instapak, instapak packaging foam, laminated copper, lamp's ballasts, lamps, loose fill, lumber, magazines, manuals, manufacturing equipment, manufacturing plastic scrap, metals, metals (scrap), microwaves, miscellaneous electronics, mixed paper, monitors, MPA tape, newsprint, office furniture, office paper, old corrugated containers, packing material (peanuts), pallets. Paper, paperboard, PBT regrind (plastic's black regrind), PET containers, phone books, photodegradable tear-away hi-cone plastic rings, pins on tape, plastic bags, plastic drums, plastic film, plastic peanuts, plastic six-pack rings, plastic: high-density polyethylene plastics (HDPE), plastic: low-density polyethylene plastics (LDPE), polycarbonate (heat proof trays), polycarbonate (multi color parts), polyethylene foam, polypropylene (PPE), polypropylene (battery trays), polypropylene (ink cartridges), polystyrene, polystyrene foam, polystyrene trays, polyurethane foam, printer cords and cables, printers, printing plates, pure water resin filters, PVC plastic trays, PVC plumbing, reusable product packaging, riceboard, rubber bands, scanners, shoes, shrink wrap, sludge from ink treatment ink, solder dross, solder scrap, solvents, steel, steel caps, stereos, stretch wrap, Styrofoam, syringe casings, televisions, tiles, tin, tin cans, toner cartridges, tooling, treatment resin filters, used oil, used oil from kitchen, used products, VCRs, wall boards, white ledger, wine barrels, wire, wood, wood pellets, wood scrap, wraps, yard debris.
Some of the products that may be composted include:
cellulose filter pads, compostable silverware, corks, floral paper, food scraps, hops, natural fiber produce bags, plate scrapings, scrap paper, seeds, soiled paper (e.g., paper napkins), spent grain, wood chips, waxed cardboard.

3.2 Paper Product Reduction Strategies

Pace at which paper use is reduced depends on the nature of the business. Internally, used of paper may be reduced through company policies, supported by employees. Externally, it takes longer time to phase out or reduce usage of paper. However, when business communicates with its customers about it going green, customers often support business initiatives and welcome changes. A list of initiatives that will allow business to reduce use of paper includes:

- Use double side copies
- Set printers to automatically duplex
- Use high-speed duplexing copying function
- Mass produced documents are automatically double-sided unless otherwise noted in client instructions
- Reduce print publications through the use of electronic media
- Make only one copy of items for distribution
- Post meeting minutes in the back room for volunteer employees without email.
- When making revisions to a document, circulate one copy with minor hand corrections rather than reprinting drafts.
- Proof and preview documents on screens before printing
- Utilize small postcards or emails, rather than large mailings
- Eliminate unnecessary forms
- Eliminate unnecessary reports
- Eliminate duplicate names from mailing lists
- Send postcards with an "opt-out" option every time.
- Encouraged employees to reuse paper when writing and printing,
- Keep reusable paper for scratch paper, notes, scatches, art work and other uses
- Place trays of reusable paper near printers and copiers
- Provide printed copies of documents to customers on request only
- Persuade the vendors and customers to use electronic signatures and make all necessary changes to documents electronically so that they only need to be printed once.
- Use recycle paper laser checks for any payments that can not be completed online

A list of initiatives that will allow business to eliminate use of paper includes:

- Automatically convert incoming faxes to emails and retire fax machine
- Make companywide effort to circulate information via electronic files, both internally and with their clients and vendors.
- Provide documents in PDF format rather than hard copy
- Generate electronic phone books
- Pay bills online.
- Replace mailings with emails and voicemails
- Send records and documents by email
- Consider multi-monitor workstations for easier proofreading without printing
- E-mail newsletters and other notifications.
- Email instead of paper forms
- Use electronics bulletin board, magazines, newsletters, etc.
- Use web pages for online access to jobs, forms, benefits and other com pany information
- Unsubscribe to all catalogs and unsolicited mailings and faxes
- Provide electronic receipts
- Provide work schedules online or through e-mails
- Send electronic invitations to company events.

3.3 Collection Stations

Collection stations are used to collect recyclable or reusable goods generated by the business, customers, employees and/or general public. Decision which of the groups of people may participate in collection depends on type of material collected, accessibility and costs. A list of goods that may be collected is extensive. Some of the possible collectible goods include:

- Phone books
- Batteries
- Cell phones (including cell phones for soldiers, shelters, etc.)
- CFL bulbs
- Compact fluorescent lights
- Computers
- Electronic equipment
- Ink cartridges
- Office supplies
- Outdated personal assistants
- Perishable food (for charities)
- Pens, pencils and markers (consider using the writing instrument brigade)
- Pet food (for shelters, Humane Society, Animal Care and Control, etc.)
- Plastic bags
- Printer toners
- Tires

3.4 Packaging

There are number of opportunities in reduction of packaging materials and costs. Since many products will arrive in a packaged form, effective strategies can be used to eliminate waste. They include:

- Reuse mailing boxes and envelopes
- Regularly reuse shipping/mailing materials (such as "peanuts")
- Collect and reuse packing peanuts from other businesses and from neighbors.
- Offer bulk sales
- Use returnable shipping containers and packaging

3.5 Source Reduction

Source reduction is the best way to eliminate bi-products. If material is not brought to the facility it does not have to be managed. This applies to both elimination and reduction of the quantity of raw materials brought in as well as innovative techniques that reduce waste. Some of the waste reduction techniques are:

- Cut hazardous chemical usage by reducing the tube diameter and length of paint and other chemical lines (less material removed when lines are cleaned)
- Replace paint with water-based (no need to extract solvents)
- Eliminate chlorinated solvents (trichloroethane) by converting solvent-based metal part cleaners to water-based washers
- Reduce building's electrical consumption
- Minimize the use of lighting fixtures by maximizing use of natural light
- Turn off computers and lights when not in use/use light switches with motion detectors
- Convert to energy-efficient lighting
- Purchase high efficiency toilet
- Re-route heat sources (e.g. Reroute heat from parts washer to the office area)
- Purchase shipping materials infrequently, only when needed
- Purchase used office furnishings
- Clean up old furniture instead of replacing
- Use materials and processes that reduce both the use of harmful chemicals and water consumption
- Phase out printers/copiers throughout their office and bring in models that use less toner
- Phase out appliances and consider energy star® units
- Encourage customers to bring and use their own shopping bags/reuse shopping bags
- Replace carpet printing process with embroidered designs where possible (large amount of water may be saved)
- Convert grass to naturalized landscape, eliminating the need to chemical fertilizer and herbicides
- Encourage employees to bring their own lunches
- Do not provide bottled water
- Use washable terrycloth fingertip towels in restrooms
- Provide re-usable plates, bowls, and cutlery
- Provide water dispenser
- Provide water fountains
- Re-use linens
- Asks clients to provide containers for their use for catering events
- Have dishware, glassware, flatware and a dishwasher available in kitchens and cafeterias to allow employees to bring in their own lunches and wash their dishes
- Have a shared dumpster
- Refill used toner cartridges
- Encourage use of refillable glass soft drinks bottles
- Use large refillable bottles for shampoos, conditioners and soaps, instead of small plastic bottles that have to be replaced every day

3.6 Food Management

Typically, no part of the used food and scraps have to go to landfill. This is an area where 100 % of the bi-products are usable. Here are some ideas of what can be done to achieve zero waste in the kitchen:

- Donate perishable food to second harvest, food banks, daycare programs, other meal programs
- Send spent grain to cattle feed
- Send used oil to bio-diesel producer or fry-oil recycler
- Consider furnace operating on used oil
- Consider alcohol recovery from fermentation bottoms
- Use waste yeast and waste beer in fuel blending
- Send meat products to rendering
- Compost cafeteria's bi-products onsite
- Consider bulk pump dispensers for condiments
- Donate excess food from events to food banks
- Remove garbage disposals from kitchens to incorporate composting as the primary means of food scrap disposal.
- Cycle fry oil through a filtration system to extend its freshness

3.7 Alternative Uses for Products

Another zero landfill strategy is reusing old products to make new. This strategy includes reusing existing products, finding new uses for products both internally and, if not possible, externally, and transforming existing products and bi-products into new products (both internally and externally). Alternative use for products is a concept of reusing old products in a new way or making new products out of existing ones. Alternative use approach goes hand-in-hand with source reduction. Many of the strategies discussed below are both sustainable and profitable. They include:

- Reuse construction materials
- Instead of installing all new carpeting, clean and replace select carpet squares
- Reuse existing structures and materials, including doors, lights, hardware, light fixtures, cabinets and plumbing products
- Send sludge generated from water treatment plant to a cement plant.
- Broadcast hops on pastures
- Use spent grain as livestock and poultry feed
- Create used fryer oil program that can provide fuel to diesel-powered vehicles, build a separate shed with heat for the mixing and storage of this bio-fuel.
- Process polystyrene into simulated wood products
- Use polyethylene foam ground in beanbag chairs
- Make reusable shopping bags from plastic mesh backs from grain
- Package compost using plastic mesh backs from grain

- Reuse or surplus furniture
- Reuse or surplus durable goods
- Consider office furniture made with recycled materials.
- Send replaced items to other offices or offer them to staff.
- Consider recycled/recyclable office supplies
- Offer refurbished computers for sale along with new models
- Donate sample products to employees
- Donate or sell old usable products
- Donate furniture
- Donate or repurpose any merchandise that gets returned to the store
- Donate house wares and hardware to Goodwill
- Donate clothes, shoes and toys to clothing banks
- Donate used construction materials to Habitat for Humanity
- Separate newspaper, bottles and cans for donation to a local religious institution
- Donate uniforms
- Donate 8 ½ × 11 sized trimming scrap to churches, schools, and non-profits
- Consider Global Soap Project, which recovers discarded soap from guest rooms and reprocesses it into new bars that are distributed to vulnerable populations throughout the world.
- Consider compostable bags for coffee grounds, which employees take home for their gardens.
- Let employees to take home compost for gardening
- Sell wood as biomass fuel and soil amendments.
- Use of sustainable nutrient cycling by turning bio-mass from gardens into mulch and compost that is then returned to garden beds.
- Use cardboard for sound reducing insulation
- Use cardboard to keep metal sheets from scratching
- Manufacture linerboard from cores from the centre of paper reels
- Make new products out of existing old ones (e.g. Make a new machine out of two broken ones)
- Consider offering free computer repair on certain days
- Sell computers to industrial shop equipment at the annual surplus sale
- Reuse or give to other shipping stores for their use shipping boxes, Styrofoam, and packaging
- Consider soy-based 'solid ink' technology to reduce packaging, waste and eliminate the ink/toner cartridges completely. Soy is fully compostable.

3.8 Green Purchases and Green Partners

Environmental solutions which involve green purchases and green partnerships improve sustainability but may have a mixed effect upon financial bottom line. Some programs may offset costs of the others. In general, care has to be used when decisions shown below are considered.

- Consider 100 % of purchases to be "environmentally green"
- Purchase packaging with at least 50 % post-consumer content
- Consider containers made of collapsible and recyclable plastic
- Consider compostable packaging (including food packaging, restaurant take-out containers, etc.)
- Consider sugarcane pulp made products (for restaurants)
- Consider post-consumer recycled content office supplies
- Use papers that are FSC-certified and contain recycled materials
- Replace toxic lead pellets with barley in a machine that manufactures orthodontic retainers.
- Use paint that is free of volatile organic compounds (low impact paint)
- Use recycled fiber carpets
- Use biodegradable carpet tiles
- Use biodegradable cleaning products
- Use biodegradable plastics
- Use plastics made with recycled materials
- Consider green partnership program with nearby school
- Offer greater selection of products made from recycled materials
- Provide customers with green services and solutions
- Survey suppliers for chemical use and EMS status
- Pressure vendors to reduce the amount of packaging materials
- Require coffee vendor to provide compostable bags for coffee grounds
- Select vendors that both offer products made from recycled materials and participate in green business practices themselves
- Ask vendors to do a joint green initiative with your partners
- Audit suppliers providing recycling and waste disposal services
- Require suppliers to design products that are durable and reusable, in factories that make dramatic reductions in air, water, and solid waste.
- Manufacture products with recovered or recycled content
- Use green building materials
- Consider recycled and rapidly renewable products during material selection, including floor coverings, steel studs and ceiling tiles.
- Put alkaline water (normally needs chemicals to treat) into fish ponds and spirulina algae
- Create a chain system, such as: chickens eat earthworms set loose in grain, spent grain is used to grow mushrooms, digester for mushroom, chicken feed and chicken wastes generates methane gas for steam for fermentation

3.9 Delivery and Transportation Systems

Improvement in sustainability of delivery and transportation (and transportation avoidance) often corresponds with cost reduction. A company may consider transportation of both employees and goods as saving and going-green opportunities. Here are some examples:

- Rebuild pallets onsite
- Rent reusable moving crates to avoid cardboard waste
- Use rented or recycled shipping pallets
- Return pallets to suppliers or reuse in-house
- Convert wood and cardboard shipping crates and boxes to reusable, recyclable and sometimes collapsible plastic containers
- Reuse in-house packing cartons from vendors
- Encourage repeat customers to return company's cartons for reuse
- Reverse-ship reusable containers to suppliers
- Incorporate delivery of recyclables into product delivery routes
- Use locally grown products, fruits and vegetables to reduce transporting emissions
- Purchase locally
- Use local vendors
- Grow portion of the served food, reuse leftovers as compost (restaurants)
- Consider having vanpools available for the use of full-time staff
- Consider employee commuter program
- Ask employees to participate in Trip Reduction Program by choosing to walk, bike, bus or carpool rather than drive themselves to work.
- Consider employee work-from-home programs
- Grind wood and green waste on site
- Require service providers (technicians, drivers, etc.) to bring recyclables back from customers
- Reuse paper packaging from received shipments to cushion outgoing products during packing
- Consider amount of solid waste in compactor to be electronically measured. It will alert both waste service/recycler and company when it is time to pick up compactor (example One Plus from Allied Waste).
- Deposit concrete and asphalt on site

3.10 Other Zero-Landfill Methods

There are many other methods that may be used. Please note that Waste To Energy is not considered a Zero Waste activity according to the Zero Waste International Alliance definition of Zero Waste. They include:

- Consider small on-site materials recovery facility ("mini-MRF")
- Consider LEED- certifying new and existing buildings
- Consider Landscaping practices such as "Zeriscape", Grasscycling
- Consider sustainable and eco-friendly practices for lawn and garden

- Consider program to eliminate hazardous waste and VOC's (usually payback for such programs is very quick due to elimination of expensive treatments, equipment and training)
- Consider onsite water treatment facility
- Develop energy management system (local Solid waste management district may help free of charge)
- Contact county recycling coordinator to take advantage of any recycling programs they have to offer.
- Consider sending factory waste to a third party for secondary segregation of recyclables
- Find ways to reduce waste during setup
- Sort recyclables after events
- Recycle coffee bags through programs like Terra-cycle Recycling Brigade
- "flip" compactor from a trash compactor to a commingled recycling compactor
- Retrofit with LED lighting
- Consider cardboard baler
- Separate materials for offsite recycling
- Consider source-separated bins in some locations, and commingled recycling bins in other locations, depending on the materials involved and nature of activities underway.
- Consider programs that takes back products after their useful life for non-landfill disposal by reusing or recycling.
- Provide quality inspection at each stage of manufacturing
- Consider zero waste logistics programs
- Consider composting shredded documents along with other food scraps
- Consider decommissioning contractor to prepare a Solid Resources Management Plan
- Consider Environmental Affairs program to eliminate potentially harmful discharges and emissions into the air, onto land, and into water

3.11 HR/PR Measures

Last but not least are Human Resource strategies that are aimed at zero waste goal that also improves bottom line and public standing of the company. Such programs demonstrate results, educate employees, customers and vendors, assign environmentally-friendly duties to employees, give opportunities for and encourage sustainable action. They include:

- Publish/promote environmental aspects
- Keep staff updated on recycling efforts and results
- Inform clients that company is going green and they consider sending them update emails

- Calculate the footprint for each office
- Monitor waste stream among staff at all locations/departments
- Share recycling results with each location/department
- Compile “Green Team”, clearly defined goals and follow through with each one
- Establish a recycling committee consisting of one or two individuals from every area
- Designate a recycling coordinator that works within the office to champion all recycling and waste reduction activities
- Consider bi-annual “green parties” to work hand-in-hand with non-profit groups on local environmental projects
- Participate in the sustainable community event
- Have sustainability initiatives supported from the top (e.g. company cars are hybrids)
- Make part of everyone’s job description is to reuse, reduce and recycle
- Orient new employees into the company’s recycling and waste reduction practices
- Periodically educate all employees about best practices for waste reduction
- Train employees in composting, recycling, source reduction
- Implement environmental employee idea boxes, employees idea surveys, online environmental chat sites for employees and/or programs like Quality Using Employee Suggestions and Teamwork (QUEST)
- Implement office recycling program in which employees have only a very small plastic bag for its trash and a larger, rigid recycling bin for its mixed paper
- Place a large recycle bins next to small trash bins
- Remove garbage bins and replace with compost and recycling bins
- Place additional recycling bins in so many locations that it would be difficult not to participate
- Place Recycling bins throughout the facility and by every garbage bin, along with guidelines detailing what is or isn’t recyclable (medical facility)
- Color code recyclable materials
- Replace the signage near waste and recycle bins to increase proper recycling well-marked recycling receptacles for glass, cans, paper, etc.
- Strongly encourage employees to reuse paper when writing and printing
- Encourage employees to eat at home or bring their own food in containers
- Educate guests about company’s recycling process, encouraging them to engage in similar activities within their own homes
- Consider recycling program and/or encouragement for visitors/guests/tenants/patients
- Consider both on-site recycling and education on the system upon move-in (tenants).
- Include helpful green tips in each (electronic) newsletter
- Send daily email with the ‘green topic of the day
- Stop using word “trash” (consider word “bi-products”)
- Give employees access to an internal classifieds program that allows the movement of durable goods from one to another (employee exchange program)

Table 2 Waste reduction possible by container—case study

Waste stream	Current monthly generation (lbs.)	Current container volume allotted (%)	Current monthly disposal Cost pertaining to stream	Potential monthly reduction (lbs.)	Potential reduction of waste stream by volume (%)	Monthly cost savings pertaining to stream
Urethane Dumpster (20 cy open top)						
Urethane flashing	2,000	22	$105.00	1,985	95	$99.75
Urethane soiled buckets	1,500	20	$95.00	1,480	95	$90.25
Urethane cans	900	20	$95.00	0	15	$14.25
Miscellaneous trash	900	18	$85.00	750	80	$68.00
OCC	300	20	$95.00	296	95	$90.25
			Monthly cost savings			$362.50
			Current monthly disposal cost of container			$480.00
			Percent savings			76 %
Pallet dumpster (30 cy open top)						
Pallets	12,000	100	$465.00	6,075	50	$232.50
			Monthly cost savings			$232.50
			Current monthly disposal cost of container			$465.00
			Percent savings			50 %
Compactor (15 cy compactor)						
Mixed plastic	1,200	50	$575.00	945	75	$431.25
Office paper	750	15	$173.00	563	75	$129.75
Label backing	2,700	15	$173.00	2,025	75	$129.75
			Monthly cost savings			$690.75
			Current monthly disposal cost of container			$1,390.00
			Percent savings			50
			Total monthly cost savings			$1,285.75
			Current monthly disposal Cost of container			$3,345.00
			Percent savings			38 %

4 Case Study

This case study is summarized from an original article published earlier this year. [1]. In March 2006, a waste assessment was performed at a light manufacturing and assembly company located in Northwest Ohio with over $25 million in annual sales. The company had been privately owned since its inception employing over 100 employees on two shifts. The company has foreign subsidiaries and ships heavily to the Pacific Rim. The objectives of the waste minimization assessment were to define the various waste streams and to identify economically feasible options for the minimization of those waste streams.

At the time of the assessment, the company had six final waste containers on site. A compactor located at the rear of the building was used for general garbage, of which approximately 85 cubic yards (cy) per month were generated. An on-call, dedicated old corrugated container (OCC) trailer located next to the compactor was collected three to four times per month. Based on the areas targeted by the company and the results of the audit, several distinct areas presented opportunities to reduce collection costs in conjunction with decreasing the amount of material which goes to the landfill. The major areas included diversion of urethane flashing and soiled buckets from the dumpster, increased diversion of OCC, aggressive office paper collection, and the diversion of mixed plastic and label backing from the compactor.

The waste assessment team identified 14 major waste streams produced by the company. Of these 14 streams, nine of the streams were potentially recyclable. At the time of assessment, the company recycled OCC, shredded office paper, pallet waste, and metal turnings; the other potentially recyclable waste streams were disposed of in either the compactor or the urethane dumpster, both of which were being landfilled. By separating the potentially recyclable material from the waste streams, the company significantly reduced both the amount of material going to the landfill as well as disposal costs. Table 2 provides the resultant alternatives for handling the recyclable materials including a cost savings analysis.

5 Conclusions

The overview of the program and case study provided in this paper demonstrates that through a cooperative effort and a creative alliance, businesses, governments, universities, and industries can work together to improve environmental quality, reduce waste, and increase company profitability. In addition, this paper provides a framework for other institutions across the world to duplicate the concepts and processes and adopt similar programs. Solid waste assessments several benefits to businesses that include:

- Cost savings and waste reduction (improved profitability)
- Improved company images (green)

Reference

1. Franchetti M (2009) The solid waste analysis and minimization research project: a collaborative economic stimulus and environmental protection initiative in Northwest Ohio. The J Solid Waste Technol Manage 25(2):88–94

Application of Hybrid MCDM Approach for Selecting the Best Tyre Recycling Process

S. Vinodh and K. Jayakrishna

Abstract

This chapter present the application of hybrid Multi Criteria Decision Making approach (MCDM) for selecting the best tyre recycling process for an Indian manufacturing organization. Selection of best recycling process is of vital importance in the contemporary manufacturing scenario which focuses on sustainability principles. Selecting the best tyre recycling process is a demanding task as it involves several criteria and attributes. Hence there exist a vital need to select the best tyre recycling process. The study presented in this chapter utilizes Fuzzy Analytical Hierarchial Process (FAHP) and Fuzzy VIKOR for selecting the best tyre recycling process. The practical aspects of the study are also discussed in this chapter.

1 Introduction

Recycling research institute defines used tyres as "An end-of-life tyre is a used tyre that cannot be reused for its originally intended purpose and is not retreaded. Such tyres may have further use as a raw material for other processes may be destined for final disposal. End-of-life tyres are also called as "scrap tyres". Used tyres are self-possessed of moderately inert material which would cause no straight damage to the environment. The intrinsic physical properties of tyres, joined with soil, garbage, gas movement, freezing and thawing, effect in the observable fact of tyre surfacing, whereby, a large percentage of buried tyres simply work their way to the surface of the landfill over a period of time. At one time, tyres were collected by waste management companies for a small fee. The tyres were classified, and fine ones were retreaded while the remaining

S. Vinodh (✉) · K. Jayakrishna
Department of Production Engineering, National Institute of Technology,
Tiruchirappalli 620 015, Tamil Nadu, India
e-mail: vinodh_sekar82@yahoo.com

J. P. Davim (ed.), *Green Manufacturing Processes and Systems*, Materials Forming, Machining and Tribology, DOI: 10.1007/978-3-642-33792-5_5,

were leftover in earth storage piles. Tyre piles are not only aesthetically unpleasant, but if handled incorrectly pose a fire hazard. Tyre fires are categorized by partial combustion resulting in substantial haze of noxious black smoke and the release of toxic oils assisting the need of choosing the best tyre recycling process. However, choosing the best tyre recycling process is a challenging task as many criteria and parameters need to be considered for this process. The conservative trial and error approach followed to choose the best tyre recycling process is often found to be inefficient. Hence a need has arised to adopt a scientific technique to choose the best tyre recycling process is realised. In this context, in this chapter the experience of adopting MCDM approach for choosing the best tyre recycling process is described.

While appling MCDM approach, FAHP was used for weighing the tyre recycling criteria in use and VIKOR (VlseKriterijumskaOptimizacija I KompromisnoResenje in Serbian, meaning multi-criteria optimization and compromise solution) was used to choose the best tyre recycling process. The methodology of VIKOR proposed by Opricovic and Tzeng [1] was used in our study. The unique advantage associated with VIKOR is that it focuses on ranking and selection from a set of alternatives in the presence of conflicting criteria [1]. VIKOR method was developed for multi-criteria optimization of complex systems. VIKOR method introduces the ranking index based on the particular measure of "closeness" to the ideal solution using linear normalization. The normalized value in VIKOR method does not depend on the evaluation unit of a criterion function. The selected recycling process was subjected to implementation in the case organization. The case study presented in this chapter indicated the practical feasibility of deploying Fuzzy based Analytical Hiearchial Process (AHP) and VIKOR for selecting the best tyre recycling process in industrial scenario.

2 Literature Review

Before carrying out the case study, the literature was reviewed from the orientations of tyre recycling processes and applications of MCDM techniques. The details of the information and knowledge gathered while conducting this literature review are presented in the following sub sections.

2.1 Review on Tyre Recycling Processes

Lebreton and Tuma [2] analyzed the present situation in both car and truck tyre markets in Germany. They applied an Original Equipment Manufacturer (OEM)-centered decision model to analyze potential future scenarios by considering their ability to raise remanufacturing rates. They conducted a real life case study in which the profitability of remanufacturing operations could be investigated in a comprehensive way. Adhikari and Maiti [3] studied about the impact of disposing waste tyres on environmental pollution. According to these authors recycle and reuse of used and waste rubber, and the reclaim of rubber raw materials are two best approaches that can be applied to prevent environmental pollution that arise due to the disposal of waste tyres. They concluded that the microwave technique is unique to get better quality product because degradation of polymer chain will not take place while applying this technique. Lia et al. [4] tested waste tyre fiber modified concrete. Concrete without and with waste tyre chips were prepared and tested as controls. A two-phase composite model was proposed for rubberized concrete. Finite element analysis was also conducted on the proposed model and the effect of various design variables on the strength of rubberized concrete. They concluded that the waste tyre fiber modified concrete has higher strength and stiffness than waste tyre chip modified concrete and the rubberized concrete was analyzed which has a much higher post-crack toughness than concrete without waste tyre rubbers. Cho et al. [5] proposed crown shape optimization using an artificial neural network (ANN) to improve tyre wear performance. In order to define the objective function, they introduced non-uniform weighting factors effectively and obtained an optimum crown contour that provides the contact pressure distribution close to an ideal one. They evaluated the central processing unit-time efficiency and reliability of optimization method using a numerical example. Sunthonpagasit and Duffey [6] observed the engineering economics of crumb rubber facilities. By conducting literature review and interviews, they developed a financial model encompassed with nominal processing operation to extent for analyzing different market, crumb size, and production scenarios. In order to improve the overall market efficiency, they suggested the analysis of the market and production impacts on financial viability in the proposed processing facilities. Appleton et al. [7] studied the use of microwave technology as an energy-efficient alternative to current heating technologies employed in the treatment and control of specific and often problematic waste-streams, including scrap tyres and plastics, and the remediation of contaminated land and groundwater. They discussed about the technical and economic factors that influence the development, application limitations associated with these technologies.

2.2 Review on Applications of MCDM Techniques

Lin et al. [8] proposed a novel hybrid MCDM technique by combining Interpretive structural modelling (ISM) and Analytic Network Process (ANP) for solving the complex and interactive vendor evaluation and selection problem. A case study was conducted in a Taiwan Semiconductor Company during which its wafer testing and outsourcing vendor was selected using the proposed MCDM technique. The results and proposed solution are applicable not only to the semiconductor companies, but also for other industries. Hambali et al. [9]

explained the application of AHP in the conceptual design stage of an automotive composite bumper beam. Using AHP, the authors selected a concept selection model namely, concurrent design concept selection and materials selection (CDCSMS) as the most appropriate selection model. Appropriate design concept was determined with the application of sensitivity analysis among eight design concepts.

Badri [10] conducted the study on service quality attributes and identified sets of quality measures. Priority weights for quality attributes were calculated using AHP, which are then, incorporated in a goal-programming model to help select the best set of quality control instruments. The author concentrated on two important issues: how to incorporate and decide upon quality control measures in a service industry, and how to incorporate AHP into the model. Opricovi [1] performed a comparative analysis of two methods namely VIKOR and Technique for Order Preference by Similarity to Ideal Solution (TOPSIS) with the aid of a numerical example. The author has illustrated the similarities between these two methods, these comparisons have been made on the basis of the showing their similarity and some differences and both based on function representing closeness to the ideal solutions. VIKOR method introduces the ranking index based on the particular measure of closeness to the ideal solution. In contrast, the basic principle of TOPSIS method is that the chosen alternative should have the shortest distance from the ideal solution and the farthest distance from the negative-ideal solution. TOPSIS method introduces two reference points, but it does not consider the relative importance of the distances from these points. Shemshadi et al. [11] used fuzzy VIKOR for carrying out the supplier selection process. These authors obtained the linguistic terms and converted them into trapezoidal fuzzy numbers. These authors used Shannon entropy concept for assigning the weights to the criteria.

2.3 Identification of Research Gap

Though researchers have concentrated on various tyre recycling processes, the application of MCDM techniques for prioritizing the best tyre recycling process is seldom investigated. Also the application of hybrid MCDM approach with respect to tyre recycling process has so for not been explored. Hence, in this study, it was decided to apply hybrid MCDM approach for prioritizing the best tyre recycling process.

3 Methodology

The study was initiated with the identification of criteria for tyre recycling process and selection of appropriate integrated MCDM methods. Based on the decision makers' inputs, weights of criteria were computed using AHP and further

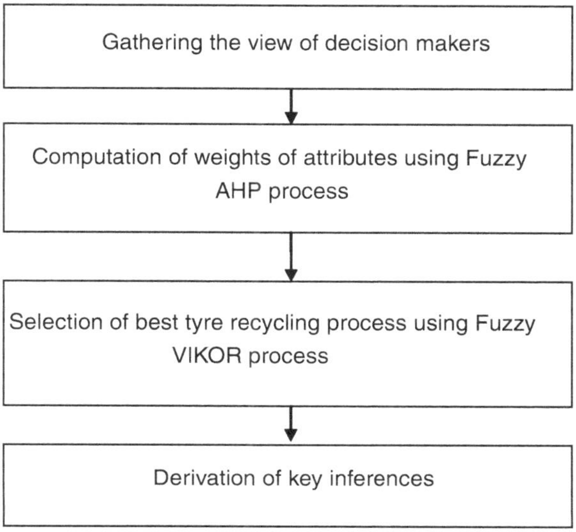

Fig. 1 Methodology adopted to conduct the case study

computations were done using Fuzzy VIKOR to identify the best tyre recycling processes. The methodology adopted in this study is shown in Fig. 1.

4 Case Study

The case study was conducted from the prespective of understanding the tyre recycling processes and prioritizing the best among them over all in terms of their possible environmental hazards they could cause to the society. A group of decision makers from various levels of organization involved in tyre remanufacturing participated in this study. These decision makers identified the elements that influence the selection of the best tyre recying proesses. These elements are described in the following subsection.

4.1 Description About the Tyre Recyling Process

The types of tyre recycling processes are briefly discussed in the following subsections.

4.1.1 Whole Tyres

In recent days, plain tyres have been largely replaced by all-weather radials. As the tread wears, these tyres lose traction on roads and motorists often replace them before they are completely worn out. Majority of these partially worn tyres are collected, shipped, and resold to the developing countries, where they are reused. Many worn but undamaged tyres are returned to the factory for retreading. The market for retreaded car tyres is relatively small.

4.1.2 Cut Tyres

In developed countries like United States of America, around 1.5 % used tyres are cut into pieces to manufacture goods such as shoe soles, gaskets, shims and blasting mats used in stone quarries [5]. This type of scrap-tyre reuse is still unknown in many parts of the globe.

4.1.3 Chipped or Shredded Tyres

Shredding is process of tyre recycling, were the used tyres are passed through a tyre shredder producing tyre strips of approximately 15–40 cm long. Tyre shreds can consume eight times less landfill space compared to whole tyres and do not resurface. Shredding plants are often built close to the supply of used tyres as shreds are cheaper to transport than whole tyres [4].

4.1.4 Tyre Crumb

Tyre crumbs are manufactured either mechanically or by cryogenic freezing process. In the mechanical process, tyres are trimmed down to chips and passed through granulators which separate loose steel and fibre and additionally rubber particle size is also being reduced. The particles thus produced are then grounded in a cracker mill to produce rubber crumb of 30–40 mesh size [12]. In the cryogenic process, tyre chips are frozen in liquid nitrogen as they pass through a cryogenic tunnel. They then pass through a series of cracker mills where they are crushed into three constituents namely rubber, steel and fabric. The cryogenic process is more expensive but it produces smoother and smaller crumbs [6].

4.1.5 Devulcanization

In the process of devulcanization, used tyre is returned to its original state as a soft, shabby, synthetic material, which are used in the fabrication of a variety of moulded or die cut rubber materials such as mats, tubs, and buckets.

4.1.6 Reclamation (Pyrolysis)

Reclamation or Pyrolysis is a thermal process that can degrade used tyres to their chemical constituents. The conventional process involves burning tyres with low oxygen content so that the tyre material is not completely converted into gases and ash. A typical automobile tyre contains approximately 4 l of oil, about 230 g of fibre, a kilogram or more of carbon black and about a kilogram of steel and methane [2].

4.1.7 Energy Recovery

The production of energy from waste is not a form of recycling but it is cost-effective as the used tyres are resold in the third world countries. On load basis, the energy content of scrap rubber is 15–20 % greater than that of coal [7].

4.2 Steps Involved in Fuzzy AHP

FAHP methodology is originally based on the concept of fuzzy set theory which was introduced by Zadeh [13]. Analysis of hierarchical structures in fuzzy environment was initially proposed by Buckley [14]. The author examined the expressions of decision makers regarding the pairwise comparisons while utilizing fuzzy ratios instead of crisp values. AHP, developed by Saaty [15] facilitates how to determine the relative importance of a set of activities in a multi-criteria decision problem. AHP makes it possible to incorporate judgements on intangible qualitative criteria alongside tangible quantitative criteria [16]. AHP proceeds by first decomposing the situation into criteria, second, comparative judgement of the criteria and third, synthesising the criteria. AHP, has been widely used to solve many complicated decision-making problems [17, 16]. The steps followed to implement Fuzzy AHP in the case company are described.

4.2.1 Arranging the Decision Making Group and Describing a Set of Relevant Attributes

Prioritizing the best tyre recycling process requires the identification of decision criteria, and establishing scales to evaluate them. These criteria must be defined according to the possible environmental effects produced by the respective tyre recycling process. A committee of five decision makers, D1, D2, D3, D4 and D5, was formed to select the best tyre recycling process. These decision makers identified the following criteria which are considered while choosing the best tyre recycling processes.

- Water eutrophication (C1)
- Air acidification (C2)
- Carbon footprint (C3)

- Total energy consumed (C4)
- Land fill (C5)
- Leachate (C6)
- Cost (C7)
- Human safety (C8)

4.2.2 Identification of Objectives of the Decision Making Process and Definition of the Problem

Decision making is the process of gathering information and selecting the optimal alternative so as to meet the decision goals. Hence, the primary step is defining the decision goal. In the case being reported here, the decision goal was to evaluate and prioritize the best tyre recycling process. After preliminary screening, eight candidate tyre recycling processes namely Whole tyres (A1), Cut tyres (A2), Chipped/shred (A3), Tyre curb (A4), Devulcanization (A5), Reclamation (A6), Plugs (A7) and Energy recovery (A8) were chosen for carrying out the evaluation. The decision makers defined the linguistic variables for the importance weight of criteria, and the fuzzy rating for alternatives with regard to each criterion so that these linguistic variables could be expressed as positive trapezoidal fuzzy numbers. Five decision makers used the linguistic weighting variables to assess the importance of the criteria.

The importance weights of various criteria and the ratings of qualitative criteria were considered as linguistic variables. In order to resolve the ambiguity and indecisiveness of human judgment, fuzzy sets theory was introduced to express the linguistic terms in decision making (DM) process. As linguistic assessments approximate the independent opinion of decision makers, linear trapezoidal membership functions were considered to be adequate for capturing the vagueness of these linguistic assessments. For example, the linguistic variable "Medium Good (MG)" can be represented as (0.5; 0.6; 0.7; 0.8), the membership function of which is given below in Eq. 1.

$$\mu_{Medium\ Good}(X) = \begin{cases} 0, & x < 0.5 \\ \frac{x-0.5}{0.6-0.5}, & 0.5 \le x \le 0.6 \\ 1 & 0.6 \le x \le 0.7 \\ \frac{x-0.8}{0.7-0.8}, & 0.7 \le x \le 0.8 \\ 0, & x > 0.8 \end{cases} \tag{1}$$

4.2.3 Conversion of Complex MCDM Problem into a Hierarchy of Interrelated Decision Elements

AHP initially breaks down a complex MCDM problem into a hierarchy of interrelated decision elements (criteria and decision alternatives). With the AHP, the objectives, criteria and alternatives are arranged in a hierarchical structure similar to a family tree. A hierarchy has at least three levels. These levels are overall goal of the problem at the top, multiple criteria that define alternatives in the middle, and decision alternatives at the bottom [18]. The hierarchy of interrelated decision elements is shown in Fig. 2.

It should be noted that the quality of the output of the AHP is strictly related to the consistency of the pair wise comparison judgments. AHP is a MCDM technique used to find the interdependency between the criteria [17, 18]. Once the problem has been decomposed and the hierarchy is constructed, prioritization procedure starts in order to determine the relative importance of the criteria within each level. The pair wise judgement starts from the second level and finishes at the lowest level, alternatives. In each level, the criteria are compared pair wise according to their levels of influence and based on the specified criteria at the higher level [18]. In AHP, multiple pair wise comparisons are based on a standardized comparison scale of nine levels. This scale is presented in Table 1.

The upper trapezoidal values were filled with values using the Saaty scale [15] and the lower trapezoidal values were generated from their inverse values. The consistency index was computed using the following Eq. 2 [18].

$$CI = \frac{\lambda_{\max} - n}{n - 1} \tag{2}$$

Here $\lambda_{\max}$ is the highest Eigen value of the resultant matrix and n represents the matrix size

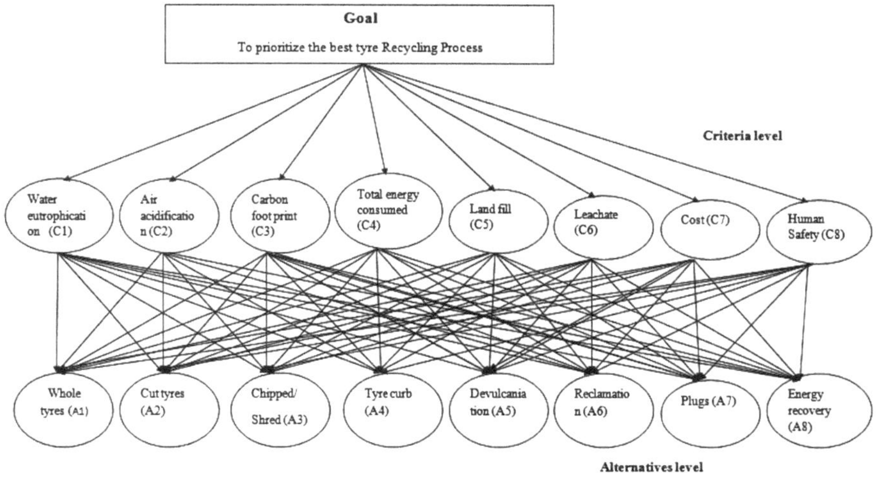

Fig. 2 Hierarchy model of interrelated decision elements

Table 1 Standardized comparison scale of nine levels

Linguistic variable	Fuzzy number
Very poor (VP)	(0.0, 0.0, 0.1, 0.2)
Poor (P)	(0.1, 0.2, 0.2, 0.3)
Medium poor (MP)	(0.2, 0.3, 0.4, 0.5)
Fair (F)	(0.4, 0.5, 0.5, 0.6)
Medium good (MG)	(0.5, 0.6, 0.7, 0.8)
Good (G)	(0.7, 0.8, 0.8, 0.9)
Very good (VG)	(0.8, 0.9, 1.0, 1.0)

The consistency ratio was computed the following Eq. 3 [18].

$$CR = \frac{CI}{RI} \tag{3}$$

where RI, Random Index which was considered as '1.24' [18]. The number 0.1 is the accepted upper limit for CR. If the final consistency ratio exceeds this value, the evaluation procedure has to be repeated to improve consistency. The measurement of consistency can be used to evaluate the consistency of decision makers as well as the consistency of overall hierarchy [19]. The linguistic variables and the corresponding fuzzy set values for each criterion is shown in Table 2.

4.3 Steps Involved in Fuzzy VIKOR

The steps followed to apply fuzzy VIKOR for prioritizing the best tyre recycling process is described in the following subsections.

4.3.1 Assortment of Input

Opricovic and Tzeng [1] developed VIKOR method in the year 2004. VIKOR determines the compromise solution. Compromise solution indicates two or more

Table 2 Excerpt of linguistic variables for tyre recycling processes

		C1	C2	C3	C4	C5	C6	C7	C8
D1	A1	VG	G	MG	VG	VG	G	VG	MG
	A2	MG	G	G	VG	G	MP	G	MG
	A3	VP	VG	MG	VP	VG	MG	VP	G
	A4	MG	G	G	G	MG	VP	MP	G
	A5	MP	MP	P	G	VG	G	MG	MG
	A6	G	MG	VP	G	MG	MP	G	VP
	A7	MG	VP	P	MP	VG	G	P	MG
	A8	P	MP	VP	MG	G	G	MG	MP

Table 3 Excerpt of linguistic variables converted into Trapezoidal fuzzy numbers

			Criteria							
			C1	C2	C3	C4	C5	C6	C7	C8
D1	Elements	A1	0.5, 0.6, 0.7, 0.8	0.7, 0.8, 0.8, 0.9	0.7, 0.8, 0.8, 0.9	0.8, 0.9, 1.0, 1.0	0.8, 0.9, 1.0, 1.0	0.7, 0.8, 0.8, 0.9	0.5, 0.6, 0.7, 0.8	0.7, 0.8, 0.8, 0.9
		A2	0.7, 0.8, 0.8, 0.9	0.5, 0.6, 0.7, 0.8	0.8, 0.9, 1.0, 1.0	0.8, 0.9, 1.0, 1.0	0.8, 0.9, 1.0, 1.0	0.8, 0.9, 1.0, 1.0	0.7, 0.8, 0.8, 0.9	0.7, 0.8, 0.8, 0.9
		A3	0.7, 0.8, 0.8, 0.9	0.7, 0.8, 0.8, 0.9	0.7, 0.8, 0.8, 0.9	0.8, 0.9, 1.0, 1.0	0.7, 0.8, 0.8, 0.9	0.8, 0.9, 1.0, 1.0	0.7, 0.8, 0.8, 0.9	0.5, 0.6, 0.7, 0.8
		A4	0.4, 0.5, 0.5, 0.6	0.4, 0.5, 0.5, 0.6	0.5, 0.6, 0.7, 0.8	0.7, 0.8, 0.8, 0.9	0.5, 0.6, 0.7, 0.8	0.5, 0.6, 0.7, 0.8	0.4, 0.5, 0.5, 0.6	0.4, 0.5, 0.5, 0.6
		A5	0.4, 0.5, 0.5, 0.6	0.5, 0.6, 0.7, 0.8	0.4, 0.5, 0.5, 0.6	0.5, 0.6, 0.7, 0.8	0.5, 0.6, 0.7, 0.8	0.4, 0.5, 0.5, 0.6	0.4, 0.5, 0.5, 0.6	0.4, 0.5, 0.5, 0.6
		A6	0.5, 0.6, 0.7, 0.8	0.4, 0.5, 0.5, 0.6	0.5, 0.6, 0.7, 0.8	0.5, 0.6, 0.7, 0.8	0.5, 0.6, 0.7, 0.8	0.7, 0.8, 0.8, 0.9	0.4, 0.5, 0.5, 0.6	0.4, 0.5, 0.5, 0.6
		A7	0.5, 0.6, 0.7, 0.8	0.4, 0.5, 0.5, 0.6	0.4, 0.5, 0.5, 0.6	0.7, 0.8, 0.8, 0.9	0.5, 0.6, 0.7, 0.8	0.4, 0.5, 0.5, 0.6	0.4, 0.5, 0.5, 0.6	0.5, 0.6, 0.7, 0.8
		A8	0.4, 0.5, 0.5, 0.6	0.4, 0.5, 0.5, 0.6	0.5, 0.6, 0.7, 0.8	0.5, 0.6, 0.7, 0.8	0.4, 0.5, 0.5, 0.6	0.5, 0.6, 0.7, 0.8	0.4, 0.5, 0.5, 0.6	0.4, 0.5, 0.5, 0.6

solutions in case of a single solution. Linguistic terms and the corresponding fuzzy numbers for each criterion and recycling processes are shown in Table 3. Here the linguistic terms and corresponding fuzzy numbers are same for both criteria and recycling alternatives weights. Fuzzy number $A \sim = (n_1, n_2, n_3, n_4)$ is said to be a trapezoidal fuzzy number if its membership function is given by Eq. 4.

$$\mu_A(x) = \begin{cases} \frac{x-n_1}{n_2-n_1}, & x \in [n_1, n_2] \\ 1, & x \in [n_2, n_3] \\ \frac{n_4-x}{n_4-n_3}, & x \in [n_3, n_4] \\ 0, & otherwise \end{cases} \tag{4}$$

where $n_1 > n_2 > n_3 > n_4$.

The importance of recycling processes with respect to the criteria was assessed by decision makers. Trapezoidal fuzzy number can deal with more uncertainty than the triangular fuzzy number [20]. It is used in modelling linear uncertainty in systematic and applied engineering problems. Let the fuzzy rating for the criterion and importance weight of the kth decision maker be X_{ijk} $\{X_{ijk1}; X_{ijk2}; X_{ijk3}; X_{ijk4}\}$ and W_{jk} $\{W_{jk1}; W_{jk2}; W_{jk3}; W_{jk4}\}$.

4.3.2 Aggregation

Since the decision variables were obtained from five different decision makers, it needs to be aggregated into a set of value. Therefore aggregation of every recycling process was carried out and the aggregated fuzzy ratings X_{ij} of alternatives with respect to each criterion was calculated using the following Eq. (5) [21]

$$X_{ij} = \{X_{ij1}; X_{ij2}; X_{ij3}; X_{ij4};\} \tag{5}$$

where,

$$X_{ij1} = \text{Min}\{X_{ijk1}\} \tag{6}$$

$$X_{ij2} = \frac{1}{k}\sum X_{ijk2} \tag{7}$$

$$X_{ij3} = \frac{1}{k}\sum X_{ijk3} \tag{8}$$

$$X_{ij4} = \text{Min}\{X_{ijk4}\} \tag{9}$$

The aggregated matrix for recycling process ratings are calculated using Eq. 5 [20] and is shown in Table 4.

Table 4 Aggregated trapezoidal fuzzy values for recycling process ratings

	C1	C2	C3	C4	C5	C6	C7	C8
A1	(0.5, 0.68, 0.74, 0.9)	(0.7, 0.8, 0.8, 0.9)	(0.7, 0.86, 0.92, 1)	(0.8, 0.9, 1, 1)	(0.7, 0.86, 0.92, 1)	(0.7, 0.86, 0.92, 1)	(0.5, 0.68, 0.74, 0.9)	(0.5, 0.76, 0.78, 0.9)
A2	(0.5, 0.72, 0.76, 0.9)	(0.5, 0.68, 0.74, 0.9)	(0.7, 0.84, 0.88, 1)	(0.5, 0.84, 0.94, 1)	(0.7, 0.86, 0.92, 1)	(0.7, 0.84, 0.88, 1)	(0.5, 0.72, 0.76, 0.9)	(0.5, 0.68, 0.74, 0.9)
A3	(0.5, 0.72, 0.76, 0.9)	(0.5, 0.72, 0.76, 0.9)	(0.7, 0.88, 0.96, 1)	(0.8, 0.9, 1, 1)	(0.7, 0.88, 0.96, 1)	(0.8, 0.9, 1, 1)	(0.5, 0.76, 0.78, 0.9)	(0.5, 0.68, 0.74, 0.9)
A4	(0.4, 0.5, 0.5, 0.6)	(0.4, 0.52, 0.54, 0.8)	(0.5, 0.64, 0.72, 0.9)	(0.7, 0.8, 0.8, 0.9)	(0.5, 0.68, 0.74, 0.9)	(0.4, 0.7, 0.72, 0.9)	(0.4, 0.52, 0.54, 0.8)	(0.4, 0.52, 0.54, 0.8)
A5	(0.4, 0.56, 0.62, 0.8)	(0.4, 0.54, 0.58, 0.8)	(0.4, 0.58, 0.6, 0.9)	(0.5, 0.64, 0.72, 0.9)	(0.4, 0.56, 0.62, 0.8)	(0.4, 0.52, 0.54, 0.8)	(0.4, 0.52, 0.54, 0.8)	(0.4, 0.52, 0.54, 0.8)
A6	(0.4, 0.56, 0.62, 0.8)	(0.4, 052, 0.54, 0.8)	(0.4, 0.58, 0.66, 0.8)	(0.5, 0.76, 0.78, 0.9)	(0.4, 0.58, 0.66, 0.8)	(0.4, 0.64, 0.66, 0.9)	(0.4, 0.54, 0.58, 0.8)	(0.4, 0.5, 0.5, 0.6)
A7	(0.4, 0.58, 0.66, 0.8)	(0.4, 0.56, 0.62, 0.8)	(0.4, 0.58, 0.66, 0.8)	(0.7, 0.8, 0.8, 0.9)	(0.4, 0.66, 0.7, 0.9)	(0.4, 0.52, 0.54, 0.8)	(0.4, 0.56, 0.62, 0.8)	(0.4, 0.54, 0.58, 0.8)
A8	(0.4, 0.52, 0.54, 0.8)	(0.4, 0.5, 0.5, 0.6)	(0.4, 0.66, 0.7, 0.9)	(0.5, 0.68, 0.74, 0.9)	(0.4, 0.56, 0.62, 0.8)	(0.4, 0.54, 0.58, 0.8)	(0.4, 0.54, 0.58, 0.8)	(0.4, 0.52, 0.54, 0.8)

Table 5 Normalised matrix

	C1	C2	C3	C4	C5	C6	C7	C8
A1	(0.56,0.76,0.82,1.00)	(0.78,0.89,0.89,1.00)	(0.70,0.86,0.92,1.00)	(0.80,0.90,1.00,1.00)	(0.70,0.86,0.92,1.00)	(0.70,0.86,0.92,1.00)	(0.56,0.76,0.82,1.00)	(0.56,0.84,0.87,1.00)
A2	(0.56,0.80,0.84,1.00)	(0.56,0.76,0.82,1.00)	(0.70,0.84,0.88,1.00)	(0.50,0.84,0.94,1.00)	(0.70,0.86,0.92,1.00)	(0.70,0.84,0.88,1.00)	(0.56,0.80,0.84,1.00)	(0.56,0.76,0.82,1.00)
A3	(0.50,0.72,0.76,0.90)	(0.50,0.72,0.76,0.90)	(0.70,0.88,0.96,1.00)	(0.80,0.90,1.00,1.00)	(0.70,0.88,0.96,1.00)	(0.80,0.90,1.00,1.00)	(0.50,0.76,0.78,0.90)	(0.50,0.68,0.74,0.90)
A4	(0.40,0.50,0.50,0.60)	(0.40,0.52,0.54,0.80)	(0.50,0.64,0.72,0.90)	(0.70,0.80,0.80,0.90)	(0.50,0.68,0.74,0.90)	(0.40,0.70,0.72,0.90)	(0.40,0.52,0.54,0.80)	(0.40,0.52,0.54,0.80)
A5	(0.40,0.56,0.62,0.80)	(0.40,0.54,0.58,0.80)	(0.40,0.58,0.60,0.90)	(0.50,0.64,0.72,0.90)	(0.40,0.56,0.62,0.80)	(0.40,0.52,0.54,0.80)	(0.40,0.52,0.54,0.80)	(0.40,0.52,0.54,0.80)
A6	(0.40,0.56,0.62,0.80)	(0.40,0.52,0.54,0.80)	(0.40,0.58,0.66,0.80)	(0.50,0.76,0.78,0.90)	(0.40,0.58,0.66,0.80)	(0.40,0.64,0.66,0.90)	(0.40,0.54,0.58,0.80)	(0.40,0.50,0.50,0.60)
A7	(0.40,0.58,0.66,0.80)	(0.40,0.56,0.62,0.80)	(0.40,0.58,0.66,0.80)	(0.70,0.80,0.80,0.90)	(0.40,0.66,0.70,0.90)	(0.40,0.52,0.54,0.80)	(0.40,0.56,0.62,0.80)	(0.40,0.54,0.58,0.80)
A8	(0.40,0.52,0.54,0.80)	(0.40,0.50,0.50,0.60)	(0.40,0.66,0.70,0.90)	(0.50,0.68,0.74,0.90)	(0.40,0.56,0.62,0.80)	(0.40,0.54,0.58,0.80)	(0.40,0.54,0.58,0.80)	(0.40,0.52,0.54,0.80)

Table 6 Crisp values for recycling process ratings

	C1	C2	C3	C4	C5	C6	C7	C8
A1	0.78	0.89	0.87	0.92	0.87	0.87	0.78	0.80
A2	0.79	0.78	0.85	0.80	0.87	0.85	0.79	0.78
A3	0.71	0.71	0.88	0.92	0.88	0.92	0.72	0.70
A4	0.50	0.58	0.69	0.80	0.70	0.67	0.58	0.58
A5	0.60	0.59	0.63	0.69	0.60	0.58	0.58	0.58
A6	0.60	0.58	0.61	0.72	0.61	0.65	0.59	0.50
A7	0.61	0.60	0.61	0.80	0.66	0.58	0.60	0.59
A8	0.58	0.50	0.66	0.70	0.60	0.59	0.59	0.58

4.3.3 Normalisation

Normalisation is carried out to have a common scale of value as it leads to have a dimensionless criteria rating. Frequently, linear normalization is employed within VIKOR method [22, 23]. The criteria with higher values are desirable and are called as positive or beneficial criteria and those criteria with smaller values are named negative or cost criteria [24]. In this technique, the criterion (C) is divided by minimum value and the benefit criterion (B) is divided by the maximum value of the decision matrix using the following Eqs. 10 and 11 [20]. The normalised values obtained using these equations are shown in Table 5.

$$u_{ij} = \left(\frac{x_{ij1}}{x_{ij4}^{+}}, \frac{x_{ij2}}{x_{ij4}^{+}}, \frac{x_{ij3}}{x_{ij4}^{+}}, \frac{x_{ij4}}{x_{ij4}^{+}}, \right), \quad C_j \in B \tag{10}$$

$$u_{ij} = \left(\frac{x_{ij1}}{x_{ij1}^{-}}, \frac{x_{ij2}}{x_{ij1}^{-}}, \frac{x_{ij3}}{x_{ij1}^{-}}, \frac{x_{ij4}}{x_{ij1}^{-}}, \right), \quad C_j \in C \tag{11}$$

where

C_j denotes the jth criterion

$$x_{ij4}^{+} = \max_i \{decision\ matrix\}, \quad C_j \in B \tag{12}$$

$$x_{ij4}^{-} = \min_i \{decision\ matrix\}, \quad C_j \in C \tag{13}$$

4.3.4 Defuzzification

In order to get crisp weights of importance of the criteria with recycling process ratings and the criteria weights, the fuzzy values were defuzzified using Eq. 14 [20, 25]. The crisp values thus obtained are shown in Table 6. Since trapezoidal fuzzy numbers were used, it could be easily defuzzified by taking the average of four values.

$$Defuzzified\ value = \frac{\mathrm{a}+\mathrm{b}+\mathrm{c}+\mathrm{d}}{4} \tag{14}$$

After getting the crisp values, the best and worst values are to be calculated. The best value (f_i^*) and worst value (f_i^-) of crisp tyre recycling alternative values were identified and are shown in Table 7.

4.3.5 Measurement of Indices

The utility (S_i), regret (R_i) and VIKOR index (Q_i) were calculated using Eqs. (15–17) [26]. The values thus obtained are shown in Table 8.

Table 7 Calculated best and worst values for recycling criteria

	C1	C2	C3	C4	C5	C6	C7	C8
f_i^*	0.79	0.89	0.88	0.92	0.88	0.92	0.79	0.80
f_i^-	0.50	0.50	0.61	0.69	0.60	0.58	0.58	0.50
	0.29	0.39	0.27	0.23	0.28	0.35	0.22	0.30

Table 8 Calculation of utility, regret and VIKOR indices

	A1	A2	A3	A4	A5	A6	A7	A8
S	0.040	0.139	0.211	0.823	0.831	0.811	0.727	0.868
R	0.016	0.068	0.095	0.353	0.238	0.238	0.224	0.263
Q	0.000	0.127	0.212	0.957	0.896	0.877	0.788	0.946

Table 9 Ranking of tyre recycling processes Recycling processes

	C1	C2	C3	C4	C5	C6	C7	C8
S	A1	A2	A3	A6	A7	A5	A4	A8
R	A1	A2	A3	A8	A5	A6	A4	A7
Q	A1	A2	A3	A8	A6	A5	A4	A7

$$S_i = \sum_{j=1}^{n} \frac{w_j \left(f_i^* - f_{ij}\right)}{\left(f_i^* - f_i^-\right)} \tag{15}$$

$$R_i = \max_i \frac{w_j \left(f_i^* - f_{ij}\right)}{\left(f_i^* - f_i^-\right)} \tag{16}$$

$$Q_i = \frac{v\,(S_i - S^*)}{S^- - S^*} + \frac{(1-v)\,(R_i - R^*)}{R^- - R^*} \tag{17}$$

Where,

Q_i, represents the ith alternative VIKOR value, $i = 1, 2, \ldots, m$;

v is used as a weight for the strategy of "the majority of criteria" (or "the maximum group utility"), whereas $1-v$ is the weight of the individual regret. The alternative having smallest VIKOR index is considered to be the best solution. Arranging S_i, R_i, Q_i in increasing order resulted in the ranking. This ranking is shown in Table 9.

Whole tyre recycling process has least VIKOR index and the solution further needed to be checked for compromise solution. During certain situations, it ended up in two or more compromise solution. Therefore it had to satisfy the conditions as stated in the next subsection.

4.3.6 Proposing Compromise Solution

This step deals with improving the alternatives for a compromise solution. The alternative ($A^{(1)}$) i.e., with highest rank by arranging S_i, R_i, Q_i in increasing order is considered to be the compromise solution if and only two conditions C1 and C2 are satisfied.

C1 *Acceptable advantage*: $Q(A^{(2)}) - Q(A^{(1)}) < \frac{1}{(m-1)}$, where $A^{(2)}$ is the second position in the alternatives ranked by Q.

C2. *Acceptable stability in decision making*: Alternative $A^{(1)}$ must also be best ranked by S or/and R. When one of the conditions is not satisfied, a set of compromise solution is selected.

The set of compromise solutions are composed of:

1. Alternatives $A^{(1)}$ and $A^{(2)}$ if only condition **C2** is not satisfied (or)
2. Alternatives $A^{(1)}$, $A^{(2)}$, ….., $A^{(m)}$ if condition **C1** is not satisfied. $A^{(m)}$ is calculated using the relation $Q\left(A^{(M)}\right) - Q\left(A^{(1)}\right) < \frac{1}{(m-1)}$ for maximum M.

In this context, C1 is not satisfied whereas C2 is satisfied; therefore the compromise solution consists of two alternatives such as $A^{(1)}$ and $A^{(2)}$. This indicates that the recycling processes namely whole tyres (C1) and cut tyres (C2) are considered to be the best solutions.

5 Results

The results obtained from Fuzzy AHP and Fuzzy VIKOR are discussed in the following sub sections.

5.1 Results from Fuzzy AHP

The weights of the criteria, which were used for evaluating the recycling processes are shown in Table 10. Table 10 indicates that the criteria C1, C2, C4 and C6 as the predominant criteria with higher environmental effects.

Table 10 Weights of criteria obtained from AHP

Sl. No.	Criteria	Weights (%)
1	Water eutrophication (C1)	35
2	Air acidification (C2)	14
3	Carbon footprint (C3)	5
4	Total energy consumed (C4)	13
5	Land fill (C5)	7
6	Leachate (C6)	10
7	Cost (C7)	8
8	Human safety (C8)	8

5.2 Results from Fuzzy VIKOR

The VIKOR index value is less for the first alternative. Therefore it could be considered as one of the best solution. While checking for the compromise solution, it indicates that condition 2 is satisfied whereas condition 1 is not satisfied. Since condition **C1** indicates that difference between first two VIKOR index values (0.127) is not greater than 0.142 (= 1/8–1). But condition **C2** is satisfied i.e. the alternative 1 is also best ranked by utility and regret indices. Therefore the compromise solution is composed of alternatives 'whole tyres (A1)' and 'cut tyres (A2)'. This indicates the recycling processes 'whole tyres (A1)' and 'cut tyres (A2)' are considered to be the best solutions.

5.3 Practical Implications

Since the application of hybrid MCDM approach in prioritizing the best tyre recycling process is found to be challenging, the study was intiated with an orientation programme with the decision makers to brief about various tyre recycling criteria and determining the recycling process. The feasibility of integrated hybrid MCDM (Fuzzy AHP-VIKOR) was explained to the decision makers for gathering necessary information required for carrying out the study being reported here. The decision makers deliberated about the essence of applying hybrid MCDM approach for selecting the best tyre recycling process. The conduct of this study indicates that the modern managers could use MCDM approach scientifically to select the best tyre recycling proces there by the End of Life disposal (EoL) aspects of tyres were taken care and there by protecting the environment.

6 Conclusion

The conduct of this study could be an initiative for prioritizing the tyre recycling process in practice across the world. The third world countries have been worst affected by the tyre recycling practices followed with low environmental concern. In this study, Fuzzy AHP and Fuzzy VIKOR was used for prioritizing the recycling process. This study also expanded the application of MCDM techniques with respect to tyre recycling process marking the uniqueness. The application of this study will better result in conceptualizing the best tyre recycling process with respect to the environmental safety concerns. The results of the study indicated that, 'whole tyre'(A1) as the best way of recyling used tyres, just by extending its life time in actual or in other forms. On the other hand, whole tyres also pollute the environment when they got piled up, but the effect is low as compared to other recycling processes.

6.1 Scope for Future Work

The case study can be extended by considering other sustainability factors like social and economic attributes of tyre recycling. This case study may also be expanded by using other MCDM techniques to validate the principles presented in this chapter.

References

1. Opricovic S, Tzeng G-H (2004) Compromise solution by MCDM methods: A comparative analysis of VIKOR and TOPSIS. Eur J Oper Res 156:445–455
2. Lebreton B, Tuma A (2006) A quantitative approach to assessing the profitability of car and truck tyre remanufacturing. Int J Prod Econ 104:639–652
3. Adhikari B, De D, Maiti S (2000) Reclamation and recycling of waste rubber. Prog Poly Sci 25:909–948
4. Lia G, Garricka G, Eggersb J, Abadieb C, Stubblefieldc MA, Pang S-S (2004) Waste tyre fiber modified concrete. Intern J Compos B 35:305–312
5. Cho JR, Shin SW, Yoo WS (2005) Crown shape optimization for enhancing tyre wear performance by ANN. Comput Struct 83:920–933
6. Sunthonpagasit N, Duffey MR (2004) Scrap tyres to crumb rubber: feasibility analysis for processing facilities. Resour Conserv Recycl 40:281–299
7. Appleton TJ, Colder RI, Kingman SW, Lowndes IS, Read AG (2005) Microwave technology for energy-efficient processing of waste. Appl Energy 81:85–113
8. Lin Y-T, Lin C-L, Yu H-C, Tzeng G-H (2010) A novel hybrid MCDM approach for outsourcing vendor selection: a case study for a semiconductor company in Taiwan. Expert Syst Appl 37:4796–4804
9. Hambali A, Sapuan SM, Ismail N, Nukman Y (2009) Application of analytical hierarchy process in the design concept selection of automotive composite bumper beam during the conceptual design stage. Sci Res Essay 4(4):198–211
10. Badri MA (2001) A combined AHP-GP model for quality control systems international. J Prod Econ 72:27–40
11. Shemshadi A, Shirazi H, Toreihi M, Tarokh MJ (2011) A fuzzy VIKOR method for supplier selection based on entropy measure for objective weighting. Expert Syst Appl 38:12160–12167
12. Chien YC, Ton S, Lee MH, Chia T, Shu HY, Wu YS (2003) Assessment of occupational health hazards in scrap-tire shredding facilities. Sci Total Environ 309(1–3):35–46
13. Zadeh LA (1975) The concept of a linguistic variable and its application to approximate reasoning. Inf Sci 8:199–249
14. Buckley JJ (1985) Fuzzy hierarchical analysis. Fuzzy Sets Syst 17:233–247
15. Saaty TL (1980) The analytic hierarchy process. McGraw Hill International, New York
16. Cheng CH, Yang KL, Hwang CL (1999) Evaluating attack helicopters by AHP based on linguistic variable weights. Eur J Oper Res 116:423–435
17. Ayag Z, Ozdemir RG (2006) A fuzzy AHP approach to evaluating machine tool alternatives. J Intell Manuf 17:179–190
18. Albayrak E, Erensal YC (2004) Using analytic hierarchy process (AHP) to improve human performance, an application of multiple criteria decision making problem. J Intell Manuf 15:491–503
19. Wang JJ, Yang DL (2007) Using a hybrid multi-criteria decision aid method for information systems outsourcing. Comput Oper Res 34:3691–3700

20. Opricovic S, Tzeng GH (2007) Extended VIKOR method in comparison with outranking methods. Eur J Oper Res 178:514–529
21. Opricovic S (2011) Fuzzy VIKOR with an application to water resources planning. Expert Syst Appl 38:12983–12990
22. Chatterjee P, Athawale VM, Chakraborty S (2009) Selection of materials using compromise ranking and outranking methods. Mater Des 30:4043–4053
23. Kaya T, Kahraman C (2011) Fuzzy multiple criteria forestry decision making based on an integrated VIKOR and AHP approach. Expert Syst Appl 38:7326–7333
24. Kuo MS, Liang GS (2011) Combining VIKOR with GRA techniques to evaluate service quality of airports under fuzzy environment. Expert Syst Appl 38:1304–1312
25. Chen SH, Hsieh CH (1999) Graded mean integration representation of generalized fuzzy number. J Chin Fuzzy Syst Assoc 5:17
26. Devi K (2011) Extension of VIKOR method in intuitionistic fuzzy environment for robot selection. Expert Syst Appl 38:14163–14168

Index

J. P. Davim (ed.), *Green Manufacturing Processes and Systems*, Materials Forming, Machining and Tribology, DOI: 10.1007/978-3-642-33792-5,